# INDUSTRIAL ROBOTS

*Computer Interfacing
and Control*

**PRENTICE-HALL INDUSTRIAL ROBOTS SERIES**

W. E. Snyder, *Series Editor*

*Snyder*, Industrial Robots: Computer Interfacing and Control

# INDUSTRIAL ROBOTS

## Computer Interfacing and Control

**WESLEY E. SNYDER**

*North Carolina State University*

PRENTICE-HALL, INC., *Englewood Cliffs, New Jersey 07632*

*Library of Congress Cataloging in Publication Data*

SNYDER, WESLEY E. (date)
    Industrial robots.

    Includes bibliographies and index.
    1. Robots, Industrial.   2. Computer
interfaces.   I. Title.
TS191.8.S67  1985        629.8'92        84-13321
ISBN 0-13-463159-5

Editor/production supervision
and interior design: **Karen Skrable Fortgang**
Manufacturing buyer: **Gordon Osbourne**

Printed in the United States of America

10   9   8   7   6   5   4   3   2   1

ISBN  0-13-463159-5  01

PRENTICE-HALL INTERNATIONAL, INC., *London*
PRENTICE-HALL OF AUSTRALIA PTY. LIMITED, *Sydney*
EDITORA PRENTICE-HALL DO BRASIL, LTDA., *Rio de Janeiro*
PRENTICE-HALL CANADA INC., *Toronto*
PRENTICE-HALL OF INDIA PRIVATE LIMITED, *New Delhi*
PRENTICE-HALL OF JAPAN, INC., *Tokyo*
PRENTICE-HALL OF SOUTHEAST ASIA PTE. LTD., *Singapore*
WHITEHALL BOOKS LIMITED, *Wellington, New Zealand*

TO ALICE, ROSALYN, AND IRENE

# CONTENTS

# PREFACE

In writing this book, I had three principal educational objectives:

To teach the basic principles of robotics and the design of robot systems.

To use the robot as motivating factor, example, and context in which to teach a number of important engineering concepts.

To collect in one volume pointers to much of the advanced literature and to explain the relevance of that literature in simple terms.

To meet my first objective, I introduce the student to a simple two-jointed ($\theta$ – $r$) manipulator. This elemental robot is used throughout the book to present the concepts of kinematics, servos, dynamics, path control, velocity control, and force control. In addition, the complete solutions for a six-degrees-of-freedom arm are provided in the chapters on kinematics.

The student who uses this book in a course will not emerge ready to design and build a sophisticated robot single-handedly, but he or she will require far less on-the-job training than was previously necessary. Furthermore, although relatively few of the students who use this text will become robot *designers*, nearly all of them will become robot *users*, many of whom will be making robot purchasing and utilization decisions. They will gain an understanding of the design, operation, and control of robots and, in addition, a knowledge of the capabilities and limitations of various types of robots which should be invaluable in that decision-making process.

Robotics is not a discipline unto itself; rather, it is an amalgam of engineering topics. As such, it is a superb context in which to teach some of those elusive topics known as *systems science*. It also provides a unifying theme in which to cover those practical engineering issues that are often reduced in emphasis in undergraduate curricula.

Thus, we unify under the one theme of robot system design selected topics from

Digital design (Chapter 2)

Electromagnetics (Chapter 3)

Controls (Chapters 5 and 9)

Motors (Chapter 4)

Mechanics (Chapter 6)

Numerical methods (Chapters 8 and 9)

Dynamics (Chapters 7, 8, and 10)

Software engineering (Chapters 5, 14, and 15)

Architectures (Chapter 14)

Languages (Chapter 15)

Instrumentation (Chapter 12)

Computer vision (Chapter 13)

In meeting my third objective, I have tried to provide simple explanations of concepts; occasionally, I have sacrificed realism in favor of simplicity. The index and reference lists are fairly complete, so that the more advanced student will be directed to state-of-the-art publications in each area. The practicing engineer will find this book to be a useful general reference for robotics concepts and a pointer to detailed literature which he or she may need.

I have used the term "engineer" several times in this preface, and the book is indeed written principally for engineers and computer scientists. The manuscript has been used three times in a senior-level electrical engineering design course at North Carolina State University, and I have found that non-EEs require at least one course in electrical engineering prior to reading Chapter 3. However, I have observed that mechanical engineers who have had the single required EE course had no problems with the Chapter 3 material, and they performed on the average slightly better than the EEs on the material in Chapters 6 through 8.

The NCSU course includes a project which adds additional relevance to the material and introduces the students to such real-world engineering problems as project group dynamics and electrical noise. Suggestions on running such a project are included, along with other suggestions for the instructor, in Notes to the Instructor.

In the interest of pedagogical soundness, I have included examples, vocabulary lists, definitions of terms and notation, and homework problems. For the same reason, there is some deliberately introduced redundancy of material.

The objective of this book is to teach concepts, not to report research results. However, developments in robotics are occurring so rapidly that references to current literature are essential. I hope that those who read this book will gain, in addition to a thorough understanding of what a robot *is*, a taste of the excitement in this young field of technology.

*Wesley E. Snyder*

**Acknowledgments**

Production of a textbook is never the work of a single person, especially not a book in a field as multi-disciplinary as robotics. I am most grateful to the following people for reviewing various pieces of the manuscript: A. Bejczy, W. Fisher, A. Goetz, W. Gruver, G. Hirzinger, F. Kauffman, R. Kelly, H. Kone, R. Paul, J. Rebman, W. Reiter, A. Sanderson, D. Stancil, and A. Yarur.

In addition, I received many useful suggestions and comments from a large number of people, including: M. Brady, S. Bonner, H. Kone, L. Sweet, W. Malpass, T. Lozano-Perèz, and G. Saridis.

I am also most grateful to my management, both at NCSU and at GE for their support and encouragement, including: S. Chitsaz, J. Erkes, S. Fix, and N. Masnari.

Of course, most of the credit goes to my wife, Rosalyn, who not only put up with my general irritability during the creation of the manuscript, but also typed the entire document (more than once), and edited it to convert what I had written into English. Her support was, and continues to, be invaluable.

Finally, I would like to thank my Rainbow personal computer for not repeatedly crashing the diskettes and blowing away entire chapters. To my TRS-80 (may it rest in peace), I can only say that it was a nice try.

# NOTES TO THE INSTRUCTOR

This book is designed to support a senior-level course for students in engineering, particularly those in electrical engineering and computer science. Prerequisites include calculus through one course in differential equations, rudimentary understanding of matrix manipulation, one undergraduate course in electronics, and one undergraduate course in digital systems.

It can also serve as a valuable reference for the practicing engineer.

The emphasis in the course is on the digital computer and its involvement in the control and operation of manipulators. From a technical point of view, it is a course about how the computer may be used to solve the relationship problems which are critical in robotics. For example, coordinate frames relate points and orientations in Cartesian space to other such points and orientations, kinematics relates the position of the hand to the angles of the joints, and vision relates perception to control. These relationships are evolved and explained through the course.

Upon successful completion of the course, the student will be prepared to make competent purchasing and utilization decisions regarding robots in manufacturing or, with minimal additional training, to be a productive member of the staff of a robot manufacturer. In the applications environment, the student will be able to assess the degree of automation required for particular manufacturing tasks, evaluate candidate robots, and if necessary, modify those robots to suit specific needs.

This course does not address the mechanical engineering aspects of robotics, such as strength of materials, mechanical time constants, or resonances. Neither does it address the aspects of power such as supplies, amplification, or control. Mention is given, however, to shielding, bypassing, and otherwise protecting digital logic from electrically noisy industrial environments.

Chapter 3 places considerable emphasis on coping with electrical noise. This chapter, more than any other, requires some background in electrical engineering. The entire chapter or selected sections may be skipped at the discretion of the instructor without serious impact on the remainder of the course.

There is more material in this book than can be covered in a one-semester course. Additional material is included to provide the student with a complete and functional reference in the fundamentals of robotics. The instructor using this as a text may choose to teach a two-semester course or may selectively exclude material. For example, electrical engineers might deemphasize much of Chapters 8 and 15; computer scientists might deemphasize Chapters 2 through 4, while placing special emphasis on Chapters 14 and 15; and mechanical engineers might place special emphasis on Chapters 8 and 10, while spending less time on Chapters 3 and 15.

## STRUCTURE OF THE COURSE

A course in robotics can be made more meaningful if it is accompanied by a project such as the one described in the next section. The project involves the design and construction of a microprocessor-based servo controller interfaced to a single manipulator joint.

Before the students begin this project, the instructor should cover the first four or five chapters of the text. The students are introduced to optical encoders, both as the recommended input device for angular position information and as a mechanism for teaching general concepts of synchronization and timing. Then, some preliminary interfacing concepts are covered, including techniques for dealing with electrical noise.

A discussion of actuators prepares the students for the presentation of the principles of control. DC motors and step motors are discussed first. The subsequent discussion of hydraulic and pneumatic actuators is intended to show the generality of the second-order differential models.

Control is presented without the Laplace transform. Only three lectures are intended for the topic of control, and only simple PE, PD, and PID methods are covered. By remaining in the time domain, the student gains a better grasp of the physical realities of mechanical system control than could be presented in such a brief period with transform techniques. The book is intended to be read and understood by students who have not had control theory.

At this point, the students are prepared to begin their projects. Specific recommendations concerning the project are given in the next section.

Once the students have begun the project work, the emphasis of the course is directed toward the computational aspects of robotics.

Mathematical representations for positions and orientations in Cartesian space are derived, initially using vectors and rotation matrices and later with homogeneous coordinates. The concept of the coordinate frame that represents both position and orientation is then presented, also using the homogeneous representation.

Using the homogeneous representation ($A$ and $T$ matrices), the kinematic equations which relate hand coordinate frames to joint angles are derived. The derivations follow closely the approaches of Richard Paul.

Kinematic velocity relationships are derived initially using differential methods, and then extended to the general Jacobian form.

At this point, the students should be able to write (non-real-time) software in a higher-level language to perform the kinematic transformations for the laboratory arm. Such an assignment must be carefully structured to avoid consuming too much student time and diverting effort from the principal project.

Trajectory control covers methods for moving the hand with respect to some externally defined (possibly moving) coordinate frame. Applications to conveyor tracking are discussed. Specific techniques include coordinate position control and joint interpolated control. An excellent supplementary project is the implementation of path control on the laboratory arm. Again, the demand on student time and effort is a consideration.

Dynamics are discussed in terms of a single joint with inertia and friction. Time-varying loads and gravity are incorporated to show the need for compensation. Physical and mathematical models for centripetal and Coriolis forces are defined, but in a limited way, for one- and two-joint arms. The difficulties of cross-coupling between joints are described and thorough studies which utilize Lagrangian mechanics are cited.

In the static case, sensing of forces and torques provides a problem environment which can easily be discussed by students at this level. Sensing of hand forces using the joint torques and the Jacobian is discussed first, and such sensing is shown to be not very effective. Then, the 3-D force/moment sensor is introduced, with a brief discussion of strain gauges.

The immediate follow up to any discussion of force sensing is compliance, and both active and passive compliance are presented and explained.

Probably the sensor with most potential for affecting the performance of robots is vision, and several lectures are intended to be devoted to this subject. The text covers the initial concepts of television and raster scanning, including references to the sampling theorem. The terms *enhancement* and *restoration* are defined to be added to the student's vocabulary, but the emphasis of this section is on image understanding.


## DESCRIPTION OF THE PROJECT

We have found that the value of the course is considerably enhanced when it incorporates a laboratory project. Groups of two or three are encouraged, since the intent of the project is to give the student first hand experience in a number of different aspects of robotics. These include

  Computer interfacing, including the ability to deal with the "nasty" parts of electrical engineering such as noise

  Use of single-chip microprocessors

  Real-time control

  Assembly language programming

These components, as well as the more computational aspects of robotics such as kinematics, will be learned through the student's work on the project.

The objective of the project is for the students to build and program a controller for a servo motor using a single-chip microprocessor. The project assumes the availability of a laboratory robot with a DC-servo motor and a optical encoder on at least one joint. There are a number of such products available; the "Rhino" is one. In the absence of a laboratory robot, a simple DC motor with encoder will suffice; however, the experience of controlling a robot joint is more effective in terms of student motivation.

In the text, interfacing and electrical noise are covered initially so that the student may begin the hardware organization of the project. The next items discussed are optical encoders and other position sensing devices. Finally, simple, single-variable control control schemes are covered. This organization enables the students to begin work on their projects about one-third of the way through the semester.

The student groups should build an interface between a single-chip microprocessor, a position sensor, and a drive mechanism. We have used both the Motorola 6801 and the Intel 8748 with some success for this application.

We have found the use of optical encoders to be advantageous, since they provide a fully digital input and eliminate the necessity for A/D converters. Furthermore, the optical encoder provides a framework in which to teach a number of digital design concepts such as timing and race conditions.

The output can be driven either from a D/A converter or by pulse-width modulation. PWM avoids the necessity of analog interfaces and provides exercise in real-time programming. We have tried using analog input and outputs and found that the students tended to consume an excessive amount of time in the analog construction without compensating learning experiences.

Each group should build a circuit board with the processor, appropriate buffer chips, bypassing, and cable connectors, to connect to the encoder/motor driver system. A mechanism must also be provided for the microprocessor to obtain input of desired position from an external source such as a terminal, other computer, or bank of switches.

The control and software aspects of the project can be graduated in levels of difficulty. PE control should be implemented first, with follow-up extensions for fast-moving groups to PD and PID or even gravity-compensated controllers.

# INDUSTRIAL ROBOTS

*Computer Interfacing
and Control*

# 1
# OVERVIEW
# OF
# ROBOTICS

This book is about robots, and about the ways in which computers are involved with robots. The book is directed primarily toward seniors in engineering or computer science.

*Industrial Robots* has two primary objectives: to teach robotics and to use robotics as an application domain in which to present many of the concepts that are important to practicing engineers and are often omitted from current curricula. These concepts include interfacing, electrical noise, synchronizing circuits, and so on.

Although we will concentrate on the industrial robots popular today, we will first briefly discuss the historical development of the concept of robotics.

## 1.1 BACKGROUND

In 1921, the Czechoslovakian playwright Karel Capek wrote a play entitled *R.U.R.* (Rossum's Universal Robots). In the Slavic languages such as Czech or Polish, the word *robot* means worker. (It may come as a surprise to a visitor to Warsaw to see posters about organizations for robots.) Today, in these countries, the word has both meanings, human and mechanical.

Capek's story, in which the mechanical servants of man revolted, became the first of a series of stories in which the robot became the "bad guy," reflecting the disenchantment with technology which was prevalent in the years following World War I. Later science fiction authors began to view the intelligent automaton in a more favorable light, again reflecting a more positive attitude toward technology by society in general. Another science fiction novel, *I ROBOT* by Isaac Asimov, published in 1950, probably best exemplifies this change.

In his novel, Asimov dealt not with the physical aspects of the machine, but with the mental processes of a machine that might be required literally to interpret and obey contradictory comments. It is interesting to note that the development of Asimov's theme was concurrent with the emergence of the computer into the public sphere—for it is the computer that has enabled the existence of robots today, and the radical decrease in cost of computation is solely responsible for the economic success and popularity of industrial manipulators.

For centuries, mechanical automata have been made by people in their own image, but when computer technology invested those automata with computational capabilities, practical applications were at last possible.

Since the computer plays such a key role in robotics, the emphasis of this book will be on the computer and what it must do to make a set of sensors, actuators, and links into a robot.

## 1.2 SOME ROBOTICS TERMS

As we use the word *robot* it has a very specific meaning. Therefore, to avoid confusion, we will now define some terms.

### Manipulator

By far, the most popular type of anthropomorphic machine is the manipulator. It most closely resembles a human arm and hand, and we will often refer to it using those terms. Virtually every machine discussed in this book is of the one-arm, one-hand type. Some development work has been done on walking machines, but emulation of legs or other parts of the anatomy is still largely in the research stage and will not be discussed here. Thus, *manipulator* means *arm*, and we will use the terms almost interchangeably.

### Teleoperator

A manipulator that operates in a tightly coupled mode with a human operator is called a *teleoperator*. The best-known example of teleoperators are the remotely controlled "hands" which run with a protective screen between them and their human operators. Such systems are specifically designed for handling radioactive materials. Today's teleoperators may have computers in their control loops to accomplish remote control or to enable the operator to sense the forces exerted by the manipulator. Such machines are theoretically capable of running autonomously, but they are not designed to operate without a human in the control loop. Therefore, the term "robot" does not apply.

### Industrial Manipulator

*Industrial manipulator* is the term normally applied to the types of robots with which we will deal. Applications of such machines are not, of course, limited to industrial tasks, and applications in the commercial and agricultural sector exist. However, by far the majority of applications of robots to date have been in industry, primarily automotive manufacturing and metalworking.

Industrial manipulators vary widely in their geometry and sensing capabilities, and in their sophistication of control. We will discuss this variety in the next sections.

## 1.3 MANIPULATOR GEOMETRIES

Manipulators come in an assortment of shapes and sizes, from the Unimate PUMA with a payload of 3 kilograms to the huge manipulator on the space shuttle, which can move 500 kg (in zero G), and heavy-duty industrial machines with payloads of over 1 ton. However, manipulators may be categorized into a fairly small set of geometric organizations that we will discuss in this section along with the impact of physical geometry on computation.

### Cartesian Geometries

Often referred to as *gantry cranes*, these robots have joints that move in rectangular orthogonal directions. These are perhaps the easiest to deal with in the computer, for a movement of $x$ inches by the $X$ actuator corresponds directly to a motion of $x$ inches by the hand. Figure 1.1 shows the geometric organization of a Cartesian robot schematically; examples of such systems in the field are shown in Figure 1.2.

### Cylindrical Geometries

In a manipulator using cylindrical geometry, the first three joints correspond directly to the three principal variables of cylindrical coordinate systems: $\theta$ (rotation), $h$ (height), $r$ (reach). The cylindrical

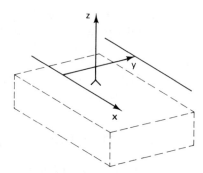

**Figure 1.1**   Rectangular coordinate system or "gantry crane" robot.

geometry is illustrated in Figures 1.3 and 1.4. If the hand position is maintained in the $\langle \theta, h, r \rangle$ coordinate system, then hand position is continuously known. Alternately, if the hand position is to be specified in $\langle x, y, z \rangle$ (as might happen if the robot is to coordinate with a conveyor), transformations must be performed relating the two coordinate systems, as illustrated in the following example.

### Example 1.1   Kinematics of a cylindrical robot

For the robot shown in Figure 1.3, determine the position of the hand in Cartesian coordinates $(x, y, z)$ as a function of the joint coordinates $(\theta, h, r)$.

**Solution:**   To solve this problem we will use geometric intuition. A more rigorous technique will be presented in Chapter 7.

The $z$ axis is the same as the vertical axis of the robot. Therefore,

$$z = h$$

The $x$-$y$ plane is perpendicular to the $z$ axis and is, therefore, parallel to the plane in which $\theta$ rotates and $r$ extends. Thus,

$$x = r \cos \theta$$

and

$$y = r \sin \theta$$

### Spherical Geometries

As shown schematically in Figure 1.5, the first three joints of these robots correspond directly to the three principal variables of a spherical coordinate system: $\theta$ (rotation), $\phi$, (rotation perpendicular to the plane of $\theta$), and $r$ (reach or radius). If the hand position is maintained in this

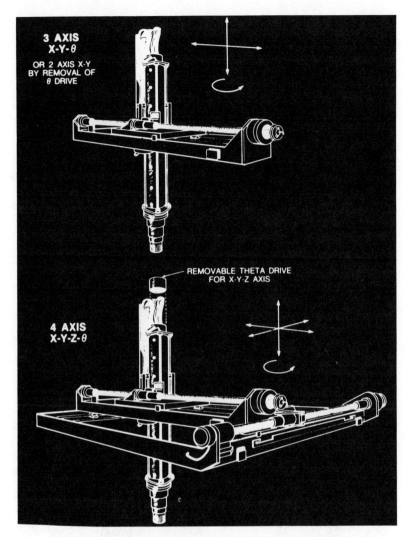

**Figure 1.2**   Gantry crane robot.  (*Courtesy General Electric*)

$\langle \theta, \phi, r \rangle$ coordinate system, then the hand position is known directly by reading the *joint variables*.    Alternately, if the hand position is to be specified as $\langle x, y, z \rangle$, then transformations must be performed to relate the hand position to the joint variables.    These transformations are known as the *kinematic transforms* and are the subject of Chapter 6. Examples of robots using a spherical geometry are shown in Figure 1.6.

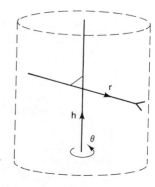

**Figure 1.3** Cylindrical coordinate system.

**Figure 1.4** Cylindrical robot. (*Courtesy PRAB*)

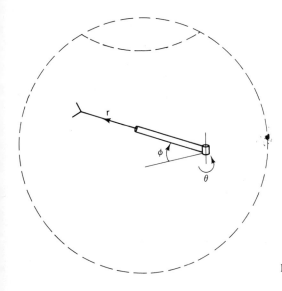

**Figure 1.5**   Spherical coordinate system.

**Figure 1.6**   Spherical robots. (*Courtesy Unimate*)

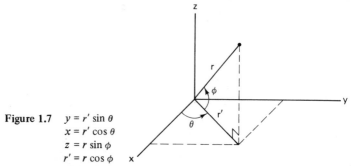

**Figure 1.7**    $y = r' \sin \theta$
$x = r' \cos \theta$
$z = r \sin \phi$
$r' = r \cos \phi$

### Example 1.2   Inverse kinematics of a spherical manipulator

For the spherical geometry of Figure 1.5, determine what the joint coordinates $r, \theta, \phi$ must be to place the hand at a particular Cartesian position, $x, y, z$.

**Solution:** As in Example 1.1, we will use geometric intuition in solving this problem. In Chapter 7, a rigorous strategy will be given that includes the extension to robots with six joints.

From the geometry shown in Figure 1.7 we determine the following relationships: we know that

$$y = r' \sin \theta, x = r' \cos \theta, z = r \sin \phi, r' = r \cos \phi$$

and

$$r^2 = x^2 + y^2 + z^2$$

Furthermore, by dividing the first two equations, we find that

$$\frac{y}{x} = \tan \theta \quad \text{and} \quad \theta = \tan^{-1} \frac{y}{x}$$

Dividing the third equation by the fourth we find that

$$\frac{z}{r'} = \tan \phi \quad \text{and} \quad \phi = \tan^{-1} \frac{z}{r'}$$

Thus we are able to find the three joint angles. The reader should note that we did not solve the third equation for $\sin \phi$ and use the inverse sine to find $\phi$. The reasons for this approach will be clarified in Chapter 7.

### Articulated Geometries

Several robots now available emulate the human elbow, introducing a rather unusual set of joint variables, consisting of three rotations for the first three joints as shown in Figure 1.8. The transformations

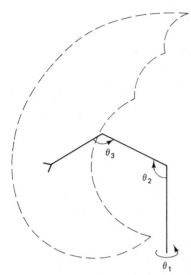

$\theta_3$

$\theta_2$

$\theta_1$    **Figure 1.8**  Articulated geometry.

from the $\langle \theta_1, \theta_2, \theta_3 \rangle$ representation to $\langle x, y, z. \rangle$ are even more complex. However, the articulated geometry is advantageous when the robot must reach over an obstacle. An example of a robot using articulated geometries is shown in Figure 1.9.

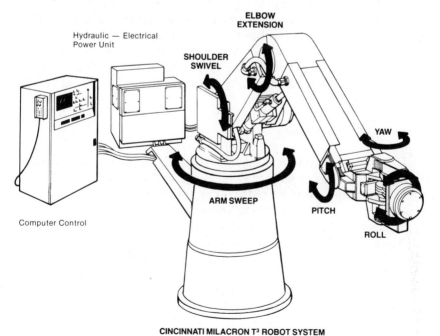

CINCINNATI MILACRON T³ ROBOT SYSTEM

**Figure 1.9**  Articulated robot. (*Courtesy Cincinnati Milacron*)

## 1.4 TYPES OF INDUSTRIAL MANIPULATORS

Industrial robots may be categorized by their geometries, as described in the last section, or by their capabilities. In this section, we will discuss the capabilities of robots in terms of types of motions that can be accomplished. We will see that the degree of flexibility of control, like most other capabilities of robots, is directly related to the sophistication of the software within the controlling computer.

### Limited Sequence: Mechanical Stop Control

The operation of mechanical stop robots is roughly analogous to that of manual typewriters and the early "programmable" sewing machines. In a manual typewriter, the tabs are set by mechanical stops. The act of pressing the tab-set key literally forces a small piece of steel out into the path of the carriage. The carriage then slams up against this piece of metal and comes to an abrupt stop, generally in the correct place. Likewise, the older programmable sewing machines were programmed by inserting a complex cam into the machine. A motor slowly rotated the cam, and the eccentricity of the cam displaced the needle the appropriate amount.

Many of the simpler robots operate in exactly the same way. That is, an actuator (typically a pneumatic piston) moves a joint until that joint runs up against a mechanical stop. Programming of such robots is typically done with a screwdriver, although some flexibility may be built in by using several selectable stops on each axis. Some experts in the field claim that mechanically sequenced machines such as these are not robots.

### Servo-Controlled Robots

If the control system is capable of detecting the instantaneous position of a joint, the actuator may be controlled in such a way that it can stop at any point along its path. It is this ability to stop at any point and not at just fixed mechanical stops that distinguishes, in many eyes, *true robots from sequenced automation.*

A servo-controlled robot can stop each of its axes at' an arbitrary point. Therefore, its hand can reach any point in its *working volume.* Given this capability, the method of programming may still vary considerably. In *point-to-point programming*, the robot controller is given commands that amount to

Go to point A, stop.
Go to point B, stop.

Open gripper.
Go to point A, stop.
Go to point B, stop.
Go to point C, stop.
Close gripper.
And so on.

In between two points, say, A and B, no control is required over either the speed of the individual axes or the path of the hand. Since the individual axes may move at different rates, or have different (angular) distances to travel, one axis may reach its destination and stop long before another. The path of the hand in such motions is completely unpredictable.

Point-to-point control is commonly used in industrial environments in which the working volume is relatively empty (and the unpredictable hand motion is unlikely to cause damage) and in which coordination with external moving objects such as conveyors is not required.

**Continuous Path Control**

In more complex work environments, the robot may be required to interact continuously with its environment. Examples include arc welding, spray painting, and performing operations along a moving conveyor. In such environments, the position and orientation of the hand must be exactly controlled all along its path. A variety of techniques exist to accomplish path control. The sophistication of the technique is dependent on such parameters as whether the path can be known in advance, as in spray painting, or whether it may vary during a motion, as in conveyor coordination. In all cases, a significant amount of data must be stored or computed.

## 1.5  COMPONENTS OF A ROBOT SYSTEM

The components of a robot system could be discussed either from a physical point of view or from a systems point of view. Physically, we would divide the system into the robot, power system, and controller (computer). Likewise, the robot itself could be partitioned anthropomorphically into base, shoulder, elbow, wrist, gripper, and tool. Most of these terms require little explanation.

Consequently, in this section we will describe the components of a robot system from the point of view of information transfer. That is, what information or signal enters the component; what logical or arithmetic operation does the component perform; and what information or signal does the component produce? It is important to note that the same physical component may perform many different information processing operations (e.g., a central computer performs many different calculations on different data). Likewise, two physically separate components may perform identical information operations (e.g., the shoulder and elbow actuators both convert signals to motion in very similar ways).

From an information transfer point of view, a robot system appears as shown in Figure 1.10. In this model, we have ignored the physical and geometrical structure of the robot and are concerned only with the information flow in the system.

### Actuator

Associated with each joint on the robot is an actuator which causes that joint to move. Typical actuators are electric motors and hydraulic cylinders. Typically, a robot system will contain six actuators, since six are required for full control of position and orientation. Many robot applications do not require this full flexibility, and consequently, robots are often built with five or fewer actuators. The properties of common actuators and techniques for driving them are discussed in Chapter 4.

### Sensor

To control an actuator, the computer must have information regarding the position and possibly the velocity of the actuator. In this context, the term *position* refers to a displacement from some arbitrary zero reference point *for that actuator*. For example, in the case of a

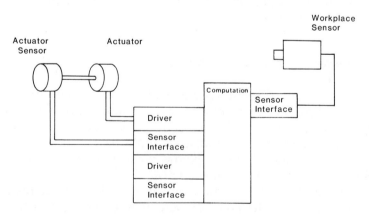

**Figure 1.10**  Components of a robot system.

rotary actuator, "position" would really be angular position and be measured in radians.

Many types of sensors can provide indications of position and velocity. The various types of sensors require different mechanisms for interfacing to the computer. In addition, the industrial use of the manipulator requires that the interface be protected from the harsh electrical environment of the factory. Sources of electrical noise such as arc welders and large motors can easily make a digital system useless unless care is taken in design and construction of the interface. This topic is covered in Chapter 3.

### Computation

We could easily have labeled the computation module of Figure 1.10 "computer," as most of the functions to be described are typically performed by digital computers. However, many of the functions may be performed in dedicated custom hardware or in arrays or networks of computers. We will, thus, discuss the computational component as if it were a simple computer, recognizing that the need for real-time control may require special equipment and that some of this equipment may even be analog, although the current trend is toward fully digital systems.

One further note: We will tend to avoid the use of the term *microprocessor* in this book and simply say *computer*, although many current robot manufacturers use one or more microprocessors in their systems. The rapid advance of VLSI (Very Large Scale Integration) technology is producing levels of integration such that for all practical purposes, every computer is a microcomputer.

The computation component performs the following operations:

**Servo (Chapter 5).**   Given the current position and/or velocity of an actuator, determine the appropriate drive signal to move that actuator toward its desired position. This operation must be performed for *each* actuator.

**Kinematics (Chapters 7 and 8).**   Given the current state of the actuators (position and velocity), determine the current state of the gripper. Conversely, given a desired state of the hand, determine the desired state for each actuator.

**Dynamics (Chapter 10).**   Given knowledge of the loads on the arm (inertia, friction, gravity, acceleration), use this information to adjust the servo operation to achieve better performance.

**Workplace sensor analysis.**   Given knowledge of the task to be performed (e.g., threading a nut), determine appropriate robot motion commands.  This may include analyzing a TV picture of the workplace (vision, Chapter 13) or measuring and compensating for forces applied at the hand (compliance, Chapter 11).

In addition to these easily identified components, there are also supervisory operations such as path planning (Chapter 9) and operator interaction.

### 1.6 OVERVIEW: WHAT IS IN THIS BOOK?

Robotics is a potpourri of topics, including geometric transforms, control theory, operating systems, motors, and signal processing.  It can be studied either for its own sake as part of the education of a robotics specialist or as an application area for a number of aspects of engineering. When studied in the second mode, it provides an important degree of breadth to the education of an engineer.

This book may be read in either mode.

In Chapters 2 and 3, issues of interfacing between robots and computers are discussed, including the interfacing of both analog and digital devices.  Illustrations are drawn from interfacing the sensors appropriate to robots; however, the principles are generalizable to most interfacing problems.  Chapter 3 is devoted to dealing with noise in digital systems, a critical concern in interfacing.

In Chapter 4, the actuators that make robots move are discussed. The more common electrical and hydraulic devices are introduced in preparation for the control discussions of Chapter 5.

Chapter 5 concentrates on control of a single joint of a manipulator.  All analysis is done in the time domain.  Proportional error, derivative feedback, and PID (proportional integral derivative) control are discussed.  Then, an extension from mathematics to software implementation introduces the concept of sampling.

Chapter 6 begins the discussion of the mathematics which are unique to robotics.  Coordinate frames are introduced as the mechanism for relating the location and orientation of one point in space to another.  Homogeneous coordinates (4 X 4 matrices) are used as the representation for frames.

In Chapter 7, the computer continues to perform transformations to resolve relationships, this time relating the location and orientation of the hand in Cartesian space to the values of the various joint angles. The set of relationships between Cartesian space and *joint space* is known as *kinematics.*

In Chapter 8, we show that kinematics not only relates the position of the hand to the (angular) positions of joints but that it can also be used to relate velocities.

Having solved the kinematic problems, we can now discuss techniques for moving the hand through paths in space. Chapter 9 discusses path, or trajectory, control.

Chapter 10 shows that kinematics is really only the steady-state solution to the problem and that, when the robot moves, dynamic equations must be considered, since new factors such as centripetal and Coriolis forces now enter the picture. The computational difficulties of finding optimal solutions are shown.

Chapter 11 introduces the need for dealing with externally applied forces, such as those inherent in threading a screw, turning a crank, or moving while pressing. Then the problems of getting a machine such as a robot to comply with such forces are presented.

More sophisticated performance by robots requires improved sensing capabilities. Chapters 12 and 13 discuss the state of the art in the fields of touch sensing and analysis and the field of industrial machine vision.

A major problem in implementation of a robot system is speed of computation. Chapter 14 addresses this issue and includes a discussion of hardware, software, and microprogramming strategies which may be used to speed up the calculation of robot control functions.

Finally, the structure of robot programming languages is presented in Chapter 15. A simple example, unloading a conveyor onto a pallet, is programmed in several of the more popular current robot languages.

## 1.7 SYNOPSIS

### Vocabulary

You should know the definition and application of the following terms:

articulated geometry

Cartesian geometry

cylindrical geometry

industrial manipulator

limited sequence robot

manipulator

path control

rectangular geometry

servo-controlled robot

spherical geometry

teleoperator

### Notation

In later chapters of this book, the symbol $\theta_i$ will be used to represent joint variables. The subscript $i$ will be larger for joints closer to the hand. In Figures 1.1, 1.3, and 1.5, we have used notation more familiar to the student. In the following definitions of symbols, we will make the notational equivalency explicit.

| Symbol | Meaning |
|---|---|
| $x$ | One of the Cartesian coordinate directions |
| $y$ | One of the Cartesian coordinate directions |
| $z$ | One of the Cartesian coordinate directions |
| $x$ (alternate use) | The first actuator of a Cartesian robot $(\theta_1)$ |
| $y$ (alternate use) | The second actuator of a Cartesian robot $(\theta_2)$ |
| $z$ (alternate use) | The third actuator of a Cartesian robot $(\theta_3)$ |
| $\theta$ | The rotary actuator of a cylindrical robot $(\theta_1)$ |
| $h$ | The vertical actuator of a cylindrical robot $(\theta_2)$ |
| $r$ | The radial actuator of a cylindrical robot $(\theta_3)$ |
| $\theta$ (alternate use) | One of the rotary actuators in a spherical robot $(\theta_1)$ |
| $\phi$ | One of the rotary actuators in a spherical robot $(\theta_2)$ |
| $r$ (alternate use) | The radial actuator in a spherical robot $(\theta_3)$ |

## 1.8 PROBLEMS

1. Figure 1.5 shows a three-degrees-of-freedom robot using a spherical coordinate system. Find the kinematic solution for this arm. That is, find expressions for the position of the hand in space $(x, y,$ and $z)$ as a function of the joint variables $r, \phi,$ and $\theta$.

2. Solve problem 1 for the articulated arm in Figure 1.8, expressing $x, y,$ and $z$ as functions of $\theta_1, \theta_2,$ and $\theta_3$.

3. Find the inverse kinematic solution for the cylindrical arm in Figure 1.3. That is, find expressions for $r, h,$ and $\theta$ as functions of $x, y,$ and $z$.

4. Solve problem 3 for the articulated arm in Figure 1.8, expressing $\theta_1, \theta_2,$ and $\theta_3$ as functions of $x, y,$ and $z$.

5. Consider the large manipulator on the space shuttle, which is used to remove satellites from the cargo bay and place them in space. Is this manipulator a robot or a teleoperator? Discuss the reasons for your answer.

6. Assume that the shuttle manipulator is a robot. Discuss how its control might differ from the control of factory robots.

# 2

---

# *SENSING POSITION AND VELOCITY*

It is rare that we can measure the position of the robot's hand directly; instead, we usually measure the position of each joint and apply the kinematic transform to find the position of the hand. Although many robots do have linear (prismatic) joints, we most often are confronted with the problem of measuring an angle, and we can, without loss of generality, discuss the measurement of angular position and extend these results to linear joints.

Measurement of position is not always necessary. For example, with step-motor–driven joints, the angular position is known simply by knowing how far the joint has been commanded to move and assuming (often fallaciously) that it really moved that far.

For most applications, however, some mechanism for sensing position is necessary. Many options exist, including using potentiometers, an inherently analog technique, and optical shaft encoders, a fully digital technique.

In the discussion of analog techniques, we will digress briefly to cover digital-to-analog and analog-to-digital conversion. This digression is made because knowledge of these techniques is essential to success in interfacing a vast variety of devices to the computer, devices which may provide the robot with the sensory capabilities it requires to deal with its environment in a sophisticated way.

Then, we will discuss the optical encoder and show that the use of the encoder provides a completely digital mechanism for determining the angular position of a joint. We will also observe that systems using the encoder suffer from the standard plagues of all digital systems: races, noise, and timing problems.

In this chapter, we will show several designs for digital circuits. The intent in describing these circuits is as follows:

1. To demonstrate design using off-the-shelf MSI (Medium-Scale Integration) chips as contrasted to design at the gate level.

2. To provide the student with techniques for coping with interfacing of nonsynchronized systems. For this reason, both races and synchronizing circuits are described.

3. To discuss the variety of options available to the designer, ranging from the well-structured sequential machine to the MSI "random logic" approach to the dedicated computer. All these techniques have different applicability in different environments. By using all three to analyze the motion of optical shaft encoders, we prepare the student to make similar logic design trade-offs in the field.

## 2.1 THE DIGITAL-TO-ANALOG CONVERTER

We assume that the reader is familiar with operational amplifiers, at least as far as using such amplifiers as analog adder circuits. Figure 2.1 shows an op-amp connected as a four-input adder. We make a small simplification of this circuit, by making all the input resistors equal to powers of two times $R_1$ and allow an input voltage of only 0 or $V_1$. The resulting unit, Figure 2.2, is a 4-bit D/A converter, where the $X_i$ are binary valued and represent the up-down position of the switches or, equivalently, the binary input to the D/A.

It is extremely difficult to fabricate such an array of resistor sizes, particularly on a single chip. (For an 8-bit D/A, the bottom resistor is

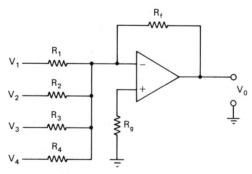

**Figure 2.1**   The circuit obeys $V_0 = -\left[\dfrac{V_1}{R_1} + \dfrac{V_2}{R_2} + \dfrac{V_3}{R_3} + \dfrac{V_4}{R_4}\right] R_f$

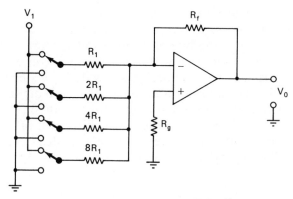

**Figure 2.2**    A 4-bit D/A converter now, $V_0 = -\dfrac{V_1 R_f}{R_1}\left[X_1 + \dfrac{X_2}{2} + \dfrac{X_3}{4} + \dfrac{X_4}{8}\right]$

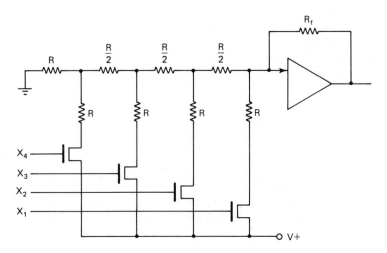

**Figure 2.3**    $R - 2R$ implementation of a D/A converter.

256 $R_1$). For this reason, the alternative D/A circuit shown in Figure 2.3 is generally used. Note also that the switches of Figure 2.2 have been realized using field-effect transistors.

Such converters are commercially available in single-chip form at very modest prices and high speeds (10 megahertz).

## 2.2 THE ANALOG-TO-DIGITAL CONVERTER

In the previous section we showed how a set of binary signals, each on a separate wire, could be composed into an analog signal on a single wire. There are many instances where, if it were possible, we would operate that device in reverse. That is, given an unknown voltage between, say, 0 and 5 V, we would produce a digital output on $n$ (typically eight) wires, such that the binary number out is representative of the input analog voltage. Of course, running the D/A converter in reverse is impossible. Instead, we must design a new device which performs the reverse function. Such a device is called an analog-to-digital converter.

In this section we will discuss only three of the many techniques of A/D conversion. The first method, which we will call linear search, is never used in commercial systems and is included here to illustrate the concept. The second two techniques are commonly found in commercial products.

### The Linear Search Converter

Figure 2.4 represents the simplest of all A/D converters. It consists of a binary counter, a D/A converter, an analog comparator, and an AND gate. The analog comparator has the property that if the voltage on the + input exceeds the voltage on the - input, the output is binary "1" (typically 5 V); otherwise, the output is binary "0" (typically 0 V).

Initially, the counter is set to a binary zero and begins incrementing as clock pulses occur. At some instant of time, the output of the D/A converter suddenly exceeds the unknown, and the comparator output switches from high to low, disabling the AND gate and stopping the counter. At that point, the number in the counter is the best digital representation of the unknown analog input.

The term *linear search* is used to represent this type of converter, since it searches from zero through every possible linearly increasing digital value until the matching value is found. On the average, an $n$-bit linear search A/D requires $2^{n-1}$ clock cycles to complete conversion.

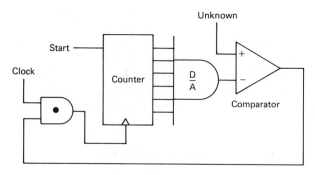

**Figure 2.4**    A simple A/D converter.

### The Successive Approximation Converter

The linear search technique of A/D conversion is analogous to searching through a dictionary looking for a definition by starting at "aardvark" and testing each word in turn to see if it matches the word being searched for. A much more appropriate technique for searching a dictionary is to open it half way and move forward or backward, according to whether the word being searched for is before or after the midpoint of the dictionary. Each time a test is made, the appropriate number of pages is cut in half. This technique is known as *binary search*, and in many applications, it is the optimal search technique in terms of time.

We can implement this strategy in an A/D converter by replacing the counter of Figure 2.4 by a *successive approximation register*, which is really a sequential machine (Figure 2.5). The strategy, which is simple, is based on manipulating the most significant bit first. The algorithm is as follows:

*Algorithm for successive approximation*

1. Let the variable $M$ represent the most significant bit.
2. Set the $M$th bit to 1.
3. If the D/A output exceeds the unknown, set the $M$th bit to 0.
4. Let the variable $M$ represent the next most significant bit.
5. If all bits have been checked, stop; else go to 2.

Figure 2.6 shows one way in which a successive approximation register could be formulated as a state diagram of a sequential machine. In this diagram, the output of the machine is represented by the low-

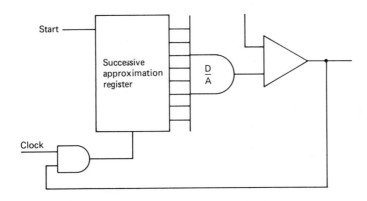

**Figure 2.5**   Successive approximation A/D converter.

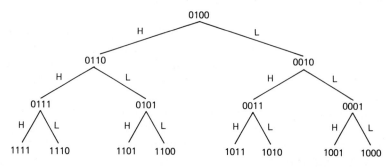

**Figure 2.6**    State diagram for a 3-bit successive approximation register.

order 3 bits of the state. The fourth bit of the state indicates completion of the conversion process. The input $H$ or $L$ represents the output of the comparator, where $H$ means that the unknown voltage is greater than the output of the D/A and $L$ means that the unknown is less than the D/A. To convert an unknown analog voltage to an $n$-bit binary representation, $n$ clock cycles are required.

**Example 2.1 Successive Approximation**

The 3-bit successive approximation register described in Figure 2.6 is used in an A/D converter to identify voltages between 0 and 8 V. The converter is presented with input of 5.6 V. For each clock cycle, show

1. State of the machine
2. Input to the comparator
3. Ouput from the comparator

**Solution**:

| Cycle | State (Prior to Clock) | Input | Output |
|-------|-----------------------|-------|--------|
| 1     | 0  1  0  0            | 4 V   | $H$    |
| 2     | 0  1  1  0            | 6 V   | $L$    |
| 3     | 0  1  0  1            | 5 V   | $H$    |
| 4     | 1  1  0  1            | 5 V   | $H$    |

**Flash Converters**

By far the fastest way in which to encode analog data into a digital form is with a *flash converter*. Such converters are fully parallel. A resistive divider network of $2^n$ resistors divides the known voltage range into that many equal increments.  A network $2^n - 1$ comparators then compares the unknown with that array of test voltages.  All comparators whose inputs exceed the unknown are "on"; all others are "off."  This *comparator code* can be converted to conventional binary by a digital priority encoder circuit.  Figure 2.7 shows a 3-bit flash encoder.

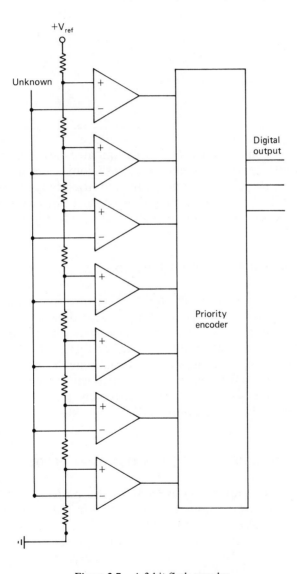

**Figure 2.7**    A 3-bit flash encoder.

**Example 2.2  Flash Converters**

The flash converter of Figure 2.7 is set up with $V_{ref} = 9$ V. An input of 6.2 V is provided. Show the state of all eight comparators.

**Solution:**  Numbering the comparators from the top of Figure 2.7, we have

| Comparator | Input on + Line | Input on − Line | Output |
|------------|-----------------|------------------|--------|
| 1 | 8 V | 6.2 V | $H$ |
| 2 | 7 V | 6.2 V | $H$ |
| 3 | 6 V | 6.2 V | $L$ |
| 4 | 5 V | 6.2 V | $L$ |
| 5 | 4 V | 6.2 V | $L$ |
| 6 | 3 V | 6.2 V | $L$ |
| 7 | 2 V | 6.2 V | $L$ |
| 8 | 1 V | 6.2 V | $L$ |

Because so many resistors and comparators are required, and because they must be precisely matched, flash converters are quite expensive. However, they are extremely fast, requiring only the settling time of the comparators plus the settling time of the priority encoder circuit. Flash converters are commonly used to encode television.

## 2.3 ANALOG MEASUREMENT OF POSITION

Possibly the most common way in which to measure angular position is with a potentiometer, with ends connected to opposite-polarity power supplies and the wiper connected to an analog-to-digital converter.

Velocity can be determined by either differentiating the potentiometer signal or by sensing an analog tachometer with the A/D converter. While using a differentiating circuit is generally far cheaper than using a tachometer, such circuits are extremely sensitive to noise; consequently, their use is not recommended.

## 2.4 DIGITAL MEASUREMENT OF POSITION AND VELOCITY

Control of a robot requires knowledge of the position and velocity of each joint. There are a number of ways in which to obtain this information. Among the more recent and more accurate techniques are those which involve a particular kind of transducer, a calibrated rotary shaft called an optical shaft encoder, and special-purpose hardware to interface the encoder to the computer.

We will describe the two major types of optical shaft encoders and present techniques that will provide signals that may be processed by a simple counter circuit for use by a computer.

### 2.4.1 Optical Shaft Encoders

There are two basic types of optical shaft encoders, absolute and incremental. The incremental shaft encoder can provide a high degree of resolution at a relatively low cost. However, in some applications the absolute optical shaft encoder is easier to use and can avoid certain problems in information transmission, which we shall explain.

**Absolute Optical Shaft Encoders.**    This type of encoder consists of a circular glass disc imprinted with rows of broken concentric arcs of opaque material.    A light source is assigned to each row with a corresponding detector on the opposite side of the disc.    The arcs and sensors are arranged so that, as the light shines through the disc, the pattern of activated sensors is a unique encoding of the position of the shaft (to within a given angular resolution).

The pattern of activated sensors is, of course, in machine-readable code. However, instead of representing the angular shaft position in the standard set of sequential binary numbers, as given by the encoder disc in Figure 2.8, absolute encoders usually employ a special Grey code, as shown in Figure 2.9.

In a Grey code, the configuration of ones and zeros is chosen so that only one bit changes between any two consecutive numbers. Grey codes are used to help minimize errors which inevitably occur when constantly changing numbers are being counted and stored. When a binary encoder is turning or a counter is counting a binary number, any or all of the bits involved may change from one reading to the next. Since

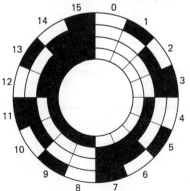

**Figure 2.8**    (Snyder 80) Absolute shaft encoder using binary code.    (*Courtesy Robotics Age*)

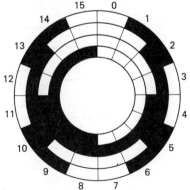

**Figure 2.9** ·  (Snyder 80) Absolute shaft encoder using Grey code.    (*Courtesy Robotics Age*)

all the bits do not change absolutely simultaneously, several transition states may occur between the true readings. In the worst cases, such as the transition from 7 (0111) to 8 (1000), every bit changes. For a few nanoseconds the system is unstable. If by ill luck the computer reads the counter or the encoder during this moment of flux, erroneous data may be recorded.

**Example 2.3  Race conditions in a binary counter**

Illustrate, using a timing diagram, a race condition that occurs when reading a binary counter or binary shaft encoder. Identify the interval of time during which a read will provide erroneous data.

**Solution:**    Consider the transition from 0111 (7) to 1000 (8), as shown in Figure 2.10 ($Y_A$ is the most significant bit). Note that a read that occurs in the center of the transition will yield a zero.

Since the duration of the instability is quite short, the probability of a bad reading is quite small, and one erroneous reading out of every thousand does not matter for most applications in which numerous readings are averaged. A motor, for instance, is much slower in its response than is an electronic controller, and it could not switch directions in response to one false signal. However, some optimal control applications may depend on near infallibility on the part of the computer (if not on the engineer), and in these applications, erroneous readings are intolerable.

Of course, digital counting circuits may be designed to avoid this problem by synchronizing their operation with a clock and reading the

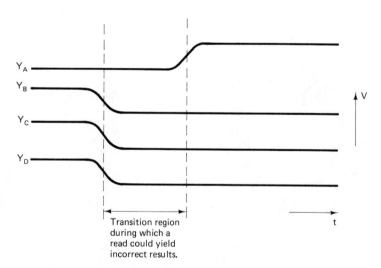

**Figure 2.10**   Race condition in a binary counter.

count only during stable intervals, but the addition of such synchronization to an absolute binary-coded encoder represents an additional expense which may be unnecessary if a Grey code is used.

Since the absolute encoder assigns to every location a unique coded number, it may be read directly by the computer, and much of the interfacing hardware that we describe here for an incremental encoder can be eliminated. With the Grey code, when erroneous readings do occur, they are much less serious, since the error is at most one count.

On the other hand, an absolute encoder requires 12 or more separate sensors to attain good resolution and is, therefore, rather expensive. An incremental shaft enroder which can identify locations within 0.04 degrees costs about $275. Others are available with much higher resolution as are less expensive models. Incremental encoders require more interfacing hardware or software, but such designs are reasonably straightforward, as we shall explain.

### Incremental Optical Shaft Encoders

Like the absolute encoder, the incremental encoder is a glass disc. However, it is imprinted with only one circular row of slots, all the same size and distance apart. One additional slot serves as a reference point.

Only three sensors are used with the incremental encoder, although the number of slots in the circular row increases depending on the resolution desired. Two of the sensors are focused on the row of slots. These are placed exactly one-half slot width apart so that, as the disc rotates, a light shining through the slots produces detector signals that are 90 degrees out of phase with each other. Figure 2.11 shows that if the disc is moving in a clockwise direction, one sensor ($V_1$) is always activated first; if the disc is moving counterclockwise, the other sensor ($V_2$) is always activated first. Thus, the pattern of the phase shift indicates direction.

The third sensor is focused on the reference point. Location can be determined by counting the number of pulses that occur from the time the reference point sensor is activated. The length of time required for a single pulse to be completed represents the velocity.

## 2.5 POSITION COUNTING HARDWARE

While all necessary information about direction, velocity, and position can be obtained directly from an incremental encoder, this information is embedded in a complex stream of pulses which is too obscure for most simple counters, since these rely on separate up or down counting signals to indicate direction. Therefore, we must provide special decod-

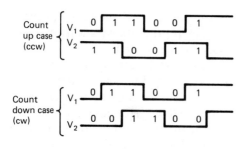

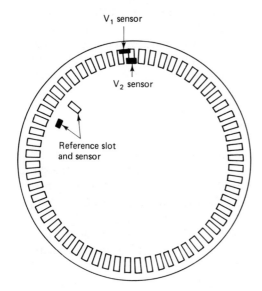

Figure 2.11   (Snyder 80) Phase rela- *
tionship of incremental encoder output.
(*Courtesy Robotics Age*)

ing logic to translate the pulses from the encoder into up and down count pulses.

It is possible to derive position simply by counting pulses directly from the encoder. One could count pulses from a single input, which, for the encoder used in our experiments, would allow us to attain an accuracy of $1/2500$ of a revolution. Even so, it is necessary to consider both inputs, since phase information is needed to determine direction and whether to count up or down.

Figure 2.11 shows the four unique states of the encoder, and the unique sequences of possible encoder outputs for continuous motion in each direction.

### 2.5.1 Decoding the Incremental Encoder Using a Sequential Machine

One could design a sequential machine to decode this complex of signals and output count up (CU) or count down (CD) pulses at appro-

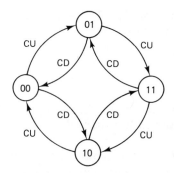

**Figure 2.12**  (Snyder 80) State diagram for a sequential machine that produces count up and count down pulses. Inputs are not shown since the next state is always equal to the current input. (*Courtesy Robotics Age*)

priate changes of the encoder.  Figure 2.12 shows such a design, where the state of the machine has been given the same designation as the "current" input.  When the input changes, so that the input does not match the current state, the machine will change into a state that does match the input and will output a CU or CD pulse as shown.

The design of a sequential machine to implement the state diagram of Figure 2.11 is reasonably straightforward.  It requires two flip-flops and a few gates.

### Example 2.4  Sequential Machine State Transition

The timing diagram in Figure 2.13 illustrates the inputs to the sequential machine of Figure 2.12. (The inputs represent a reversal in direction of an incremental shaft encoder.)  Also shown is the clock for that machine.  Sketch on the same axes the state of the machine.

**Solution:**  From Figure 2.12, we recognize that the state is represented by two bits and, therefore, is stored by two flip-flops.  We denote the outputs of these flip-flops as $Y_A$ and $Y_B$.  Note that the relationship between clock frequency and

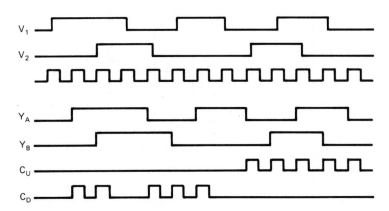

**Figure 2.13**   Timing of sequential machine state.

frequency of the encoder pulses in Figure 2.13 is not realistic. To ensure proper operation of a machine of this type, the clock should be much faster.

### 2.5.2 Decoding the Incremental Encoder Using Synchronizing Flip-Flops

Alternatively, one could sacrifice resolution for simplicity and get a design which counts every cycle of the encoder rather than every state. Figure 2.14 shows one such design.

Special attention should be paid to the design of Figure 2.14, for it is based on a pair of very general and very useful synchronizing circuits. Consider the pair of flip-flops, $ff_1$ and $ff_2$. They are rising edge–triggered flip-flops (SN7474). That is, they latch the data on the $D$ input at the instant of a rising edge on the clock signal.

Furthermore, they have an asynchronous, overriding clear input

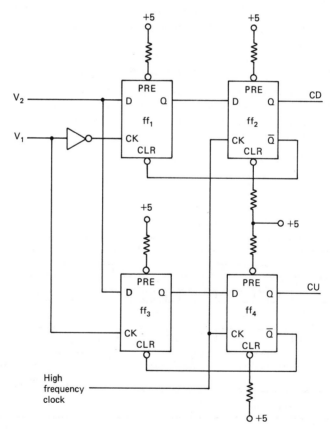

**Figure 2.14**   (Snyder 80)  A low-resolution encoder. (*Courtesy Robotics Age*)

which is asserted when low. The interconnection shown provides the characteristic that when a single event (falling edge) occurs on the $V_1$ line, a single pulse will be output from $ff_2$ and, furthermore, that pulse will be synchronous with the clock signal. The student can easily show this.

**Example 2.5  A Synchronizing Circuit**

Figure 2.15(a) shows a variation of the basic synchronizing circuit of Figure 2.14. Given input and clock signals as shown in Figure 2.15(b), show the output of the $ff_1$ and $ff_2$.

**Solution:**    See Figure 2.15(c).

One final alternative technique for decoding the incremental encoder output makes use of a feedback ROM (read only memory). That

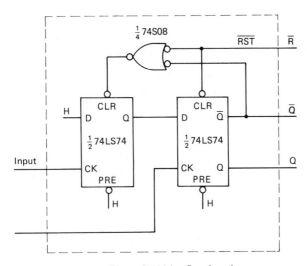

**Figure 2.15(a)**    Synchronizer.

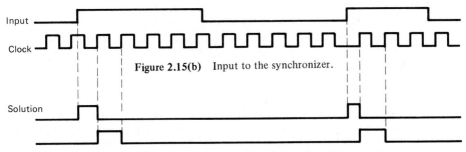

**Figure 2.15(b)**    Input to the synchronizer.

**Figure 2.15(c)**    Output from the synchronizer.

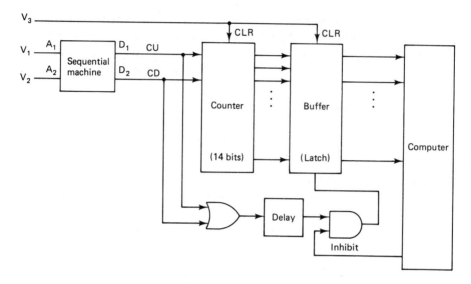

**Figure 2.16** (Snyder 80) Block diagram of position hardware. (*Courtesy Robotics Age*)

design requires only one chip, but it is dependent on critical timing conditions. The details of the ROM-based design can be found in Snyder and Schott (1980).

We have thus developed a system which decodes the direction and position information from an incremental shaft encoder. Of course, making use of that information requires more chips, a counter, and a buffer memory, as shown in Figure 2.16. In that figure, the sequential machine of Figure 2.12 provides CU and CD pulses to a counter. The counter then keeps track of the position of the shaft encoder. Figure 2.16 also includes an interface to a computer and uses a latch to avoid a timing/synchronization problem. This problem results from the possibility that the computer may read the counter while it is still changing.

One way of eliminating this timing conflict is to provide an output from the computer which indicates *computer about to read,* which will stop all counting activity. But, if a count pulse occurs during this time, it will be missed. To avoid missing any pulses, an additional level of memory must be used which will read (in parallel) the contents of the counter each time it is updated and hold that value for the computer. The "computer about to read" signal will then inhibit (we will call this signal the INHIBIT from now on) the loading of this memory.*

---

*One might also consider parallel peripheral circuits which are commonly used with 8-bit microprocessors and will soon be available in the 16-bit field. For example, the Intel 8255 can, in some designs, provide a similar latching function to that provided by the buffer in this design (latching only on request).

The count up and count down pulses operate as described on the counter, and their logical OR is taken, delayed (to let the counters settle), and used to strobe the latch. When the computer is about to read, the inhibit signal (normally 1) will be set to 0, preventing the occurrence of the strobe to the memory. After reading, the computer sets INHIBIT back to 1, allowing the latch to be loaded once again.

### 2.5.2 Design Considerations

An engineer will often choose to modify a design, either to meet the requirements of a specific circumstance or to improve efficiency. However, when dealing with encoders, it is crucial to note that it is possible to receive arbitrarily short pulses from the encoder due to instantaneous direction reversals. These may be induced by mechanical jitter and/or motion around zero speed. Therefore, it is essential that no pulses be permitted whose duration is less than the shortest allowable clock period in the electronic counting logic. This may be easily accomplished by using a low-pass filter followed by a Schmitt trigger on the encoder outputs. In the next chapter, this circuit will be discussed in detail.

## 2.6 HARDWARE FOR MEASURING VELOCITY

Since velocity is equal to distance divided by time, there are at least two ways in which to measure it. One could measure how many count pulses occur in a specific time interval, or one could measure the amount of time required by one encoder pulse. There are advantages and disadvantages to both approaches.

In the first case, one must choose a time interval sufficiently long to accumulate a number of pulses, to determine velocity accurately. For example, suppose that the maximum velocity of the joint is 1 rev/sec (revolution per second). Then, at this rate, our 10,000-count encoder yields 1 count pulse every 100 $\mu$s (microseconds). For an 8-bit accuracy, we should allow time sufficient to accumulate 255 counts, or 25.5 ms (milliseconds). This gives us a new velocity measurement approximately 40 times a second, an update rate which is marginally fast enough for a small, fast robot, but quite sufficient for larger, more massive machines.

On the other hand, one could measure the length of a single encoder pulse. This yields an update rate which is quite high but, at the same time, is rather noise sensitive. In this section, we will discuss this second approach, despite its noise sensitivity. We will also explain a method for dealing with noise. At the conclusion of this chapter, we will return to the issue of which method to use and propose yet another approach, one that utilizes a computer.

The basic idea behind our velocity hardware is the measurement of the width of a single cycle from the encoder. This is accomplished by first dividing the frequency of $V_1$ by 2 using a simple flip-flop* producing the signal $V_1'$ and then using a clock whose frequency is much higher than the maximum frequency of $V_1'$ and counting the number of clock pulses that occur while $V_1'$ is low. Again, assuming a 2500-count/rev encoder, with a maximum velocity of 1 rev/sec, we find that the shortest period which $V_1'$ can be low is 400-$\mu$s. A 2.5-kHz (kilohertz) clock will give 1 count at this rate.

If we choose to use a 2.5-kHz clock, 255 counts will require 102 ms. Consequently, an 8-bit counter will overflow if it is clocked at 2.5 kHz and the $V_1$ pulse is high for more than 102 ms. This corresponds to a velocity of 1 rotation every 4 minutes. Thus, with a 2.5-kHz clock and an 8-bit counter, we can measure any velocity from 1 rev/sec down to 1 revolution every 4 min (minutes). We assume that higher velocities are impossible and that slower velocities are stationary. With this system, the accuracy goes down at higher velocities, with a worst case error of 25 percent at the maximum speed. While this is the worst case error on any sample, the 2.5-kHz clock is uncorrelated with the encoder, and measurement errors will tend to average out, even over a few measurements. Of course, at lower velocities, errors are much smaller.

### Example 2.6  Velocity Determination

The velocity counter circuit of Figure 2.17 is used to determine the frequency of rotation of a shaft. The shaft is instrumented with an incremental shaft encoder having 1000 counts/rev.

> Given that the shaft is turning at 2 rad(radians)/sec, determine
> 1. The duration of the $V_1'$ pulse
> 2. The maximum number to which the counter will count before being cleared
> 3. An estimate of the accuracy of the measurement

**Solution:**

1. The shaft is turning at 2 rad/sec, or 0.318 rev/sec. A 1000-count/rev encoder thus undergoes 318 counts/sec when turning at this rate. The width of the $V_1'$ pulse is equal to the period of the $V_1$ signal, or 1/318 = 3.14 ms.

2. Using a clock of 2.5-kHz, the counter will be incremented every 1/2500 = 400-$\mu$s. Thus, in 3.14 ms, the counter will be incremented approximately

$$\frac{3.14 \times 10^{-3}}{4 \times 10^{-4}} = 7.85 \text{ times.}$$

---

*We time an entire cycle of $V_1$ rather than simply the time $V_1$ is high, nontemporal since the duration of a cycle is more repeatable than is the high or low period.

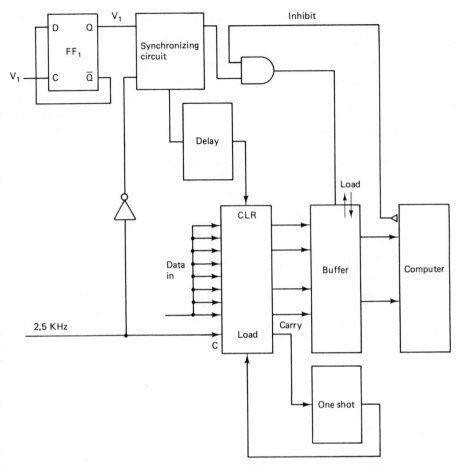

**Figure 2.17**    (Snyder 80) Hardware for measuring velocity. (*Courtesy Robotics Age*)

Of course, the counter cannot be incremented a fractional number of times. It will in fact be incremented only seven times.

3.  A count of 7 corresponds to a duration of $7 \times 400 \ \mu s = 2.8$ ms. At this rate, the encoder would require 2.8 sec for 1000 counts, for a velocity of

$$\frac{1}{2.8 \ \text{rev/sec}} = 2.24 \ \text{rad/sec}$$

The true velocity was given as 2 rad/sec. The estimate is thus in error by 0.24/2.24 = 11%.

Figure 2.17 shows a block diagram of a "quick and dirty" solution to the velocity measurement electronics. Counting occurs while $V_1'$ is low. That is, we time the "off" half of the $V_1'$ pulse.

As long as $V_1'$ is low, the counter is enabled for counting. Every 400 $\mu$s, the counter will be incremented by 1. As soon as the rising edge occurs on $V_1'$, the memory will be strobed (a rising-edge triggered latch is used), retaining the contents of the counter at that moment, and then, after the delay, the counter will be cleared. The clear to the counter is an overriding clear, and the counter will stay cleared, even though clock pulses occur, until $V_1'$ goes low again.

Note that, to avoid the possibility of a load signal to the latch occurring while the counter's outputs are changing, $V_1'$ must be synchronized with the clock. This is accomplished by the addition of a second flip-flop, whose synchronized output is used instead of the original $V_1'$. Also, as in the position circuit, the load pulse to the latch is gated by the computer INHIBIT signal.

The only remaining difficulty to be handled is the possibility that, at low velocities, the $V_1'$ pulse is so long that the counter overflows. In this case, a carry output from the counter occurs, triggering a one-shot,* which loads the counter with all 1's. At the next cycle of the clock, another carry will again trigger the one-shot and again load all 1's. Consequently the counter will always contain hex FF when the arm is stationary.

## 2.7 DETERMINING DIRECTION

The direction of motion (i.e., the sign of the velocity) needs to be considered at this time. It is vital to remember the distinction between velocity and speed, between vector and scalar quantities. We will make use of velocity informaton in our study of control systems, and the proper sign of the velocity will be essential to proper control.

The circuit up to this point has considered only the width of the encoder pulse, corresponding only to speed and not to direction. We have handled the question of direction by using one additional flip-flop and using the CU and CD pulses from the position counting hardware to set or clear, respectively, this bit. There is no race condition to be considered when interfacing this bit to the computer, since erroneous readings could only occur at zero (or very low) velocities, and in that case, it does not matter.

---

*The author normally advises against designing with conventional one-shots, which are particularly sensitive to noise and prone to false triggering. Instead, the synchronous circuit of Figure 2.14 should be used. This functions in an analogous way to a one-shot without the noise sensitivity.

## 2.8 ESTIMATING ACTUAL VELOCITY
## FROM THE MEASUREMENT

In designing our velocity hardware, we made a trade-off between the number of bits in our velocity counter and the accuracy with which we could determine velocity. Also entering our calculations were the maximum and minimum measurable speeds. Since we arbitrarily determined to use an 8-bit counter, we introduced significant errors at the high end of the velocity range. These errors could be reduced by using more bits and a higher clock rate, thus measuring higher velocities more accurately, or by continuing to use 8 bits, but with a faster clock, setting a higher minimum speed.

A third alternative is to anticipate these errors and to allow them to average out. We have chosen this alternative by using a simple low-pass filter implemented in software. If $V_{est}$ is the velocity to be estimated and $V_m$ is the measured velocity, we compute $V_{est}$ by

$$V_{est} = 0.5 \ V_{est} + 0.5 \ V_m$$

Other choices of coefficients are possible, corresponding to more or less averaging, but we have found this simple filter to be quite satisfactory in our experiments.

The quantity measured by the hardware of Figure 2.17 is a measure of time, not velocity. To convert to velocity, one must divide a constant by the measured number. One could consider doing this division by a lookup table, utilizing a ROM. However, the $V = 1/T$ calculation has a hyperbolic graph, which at low velocities results in a large number of different time periods, all of which round off or truncate to the same values. Thus, by choosing to use an 8-bit lookup, we would have introduced errors at the low-velocity end of the scale due to rounding off. We were already inaccurate at high velocities, and more inaccuracy was clearly undesirable. For this reason, we elected to perform the division in the computer.

In fact, we have found it most effective to do the previously described averaging/filtering operation on the measured data prior to performing the division, using 16-bit arithmetic on the computer.

## 2.9 CONCLUSION AND A DESIGN ALTERNATIVE

We have described special-purpose hardware for determining position and velocity using data from an incremental shaft encoder and for presenting this information to a computer, taking into account the possibility of timing conflicts during reading. We have also included some concepts that are generalizable to a large class of digital hardware problems, including the rather standard but useful two–flip-flop syn-chronizing circuit (Figure 2.14).

Having provided this information, the best advice we can give is: Don't take our advice blindly! That is, it is probably not necessary to use special hardware at all. We have often observed that, in a holdover from the time when computers were very expensive, engineers will sometimes interface to a microprocessor using hardware that is more expensive than the microprocessor itself.

The most effective, easiest to use, and most inexpensive hardware you can use is probably a computer. Both the algorithms that we have described in terms of special hardware, position hardware, and velocity hardware are easily implemented in software, using interrupts. Many computer manufacturers provide I/O (input/output) chips that selec-tively interrupt on either the rising or falling edge of the input signal. The Motorola MC6820 is one such example. With such a device, the sequential machine of Figure 2.12 can be implemented in simple software.

Furthermore, with single-chip computers like the Intel 8748, the total amount of required hardware may be one chip.

Velocity determination may likewise be performed by a single-chip computer with an internal timer. Yes, use two computers, one for velocity and one for position! The cost is so low that the dedication of a single-chip processor may very well be the most effective means to implement such a system.

## 2.10 SYNOPSIS

### Vocabulary

You should know the definition and application of the following terms:

absolute optical shaft encoder
comparator
digital-to-analog converter
edge-triggered flip-flop
filter
flash encoder
Grey code
incremental encoder
inhibit
interrupt

linear search converter
operational amplifier adder
phase
successive approximation
synchronizer
tachometer

## 2.11 REFERENCE

SNYDER, W. E., and SCHOTT, G. Using optical shaft encoders, *Robotics Age,* Fall 1980.

## 2.12 PROBLEMS

1. Figure 2.12 shows the state diagram for a sequential machine that generates the count up and count down pulses for an up-down counter from the quadrature outputs of an incremental shaft encoder. Implement this design using $J$–$K$ flip-flops.

2. Repeat problem 1 using $D$ flip-flops. Was this design any easier? Why?

3. Assume that a sequence of pulses exactly like those shown in Figure 2.11 (count up case) are input to the circuit of Figure 2.14. Assume that the pulses are 500 ms wide. Furthermore, assume that the clock input is a 50-kHz clock. Construct a detailed timing diagram showing the signals on the $Q$ outputs of all four flip-flops.

4. Figure 2.12 shows the design of a simple circuit for a quadrature decoder. This could be done in software. Assume that you have a computer with an I/O port which can be set to interrupt on rising or falling edges of pulses. That is, two separate inputs exist with the property that if a rising/falling edge occurs on that input, an interrupt will occur. Furthermore, assume that the CPU has some means for knowing which input caused the interrupt. Develop and flow chart an algorithm that keeps track of position in a software counter and uses no other chips.

5. Assuming that you are using a SN74193 binary up-down counter to keep track of position, design a circuit utilizing SN7474 edge-triggered $D$ flip-flops and SN7400 NAND gates, which will accept the $V_1$ and $V_2$ outputs from an incremental counter and generate the appropriate pulses to the 74193. The counter should count up (or down, as appropriate) once for each cycle of the $A$ line. You may make use of Figure 2.14 as part of your design.

6. You have some optical shaft encoders with 1000 cycles of the $V_1$ output for one revolution of the shaft. Assuming maximum speed of the shaft is one revolution per second, discuss the required clock speeds.

7. On the timing diagram in Figure 2.P.1, show the outputs of the four flip-flops in Figure 2.14.

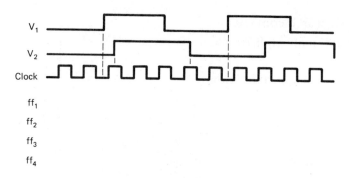

Figure 2.P.1

8. Design a successive approximation register (sequential machine) for a 4-bit A/D converter. A state diagram is all that is required.

9. Flow chart a program to perform A/D conversion using the successive approximation algorithm. Assume that the computer has one input, from the analog comparator, and eight outputs, to the D/A.

10. Design a circuit which has one signal input, $A$, and one output, $B$, plus a high-frequency clock input. Each time a pulse occurs on $A$, one and only one pulse should occur on $B$, and furthermore, those pulses should have the timing relation shown in Figure 2.P.2.

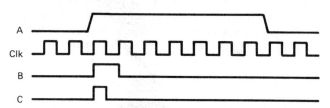

Figure 2.P.2

11. It might be desirable to produce a signal on $B$ which is high only when the clock is high. One way to do this would be to AND B with the clock. Would this work in your design for the solution of problem 10? If not, explain why. How could you produce such a pulse?

# 3

---

# *NOISE IN DIGITAL CIRCUITS*

In Chapter 2, we discussed building special hardware to determine position and velocity from optical shaft encoders. We also described how to perform these same functions in software by connecting the encoders directly to a microprocessor. If you were to take a wire-wrap tool and build the circuits described in that chapter, they might work in the laboratory. However, they most certainly would not work on the factory floor; not unless you have taken precautions to deal with electrical noise. Before we attempt to characterize noise or learn how to deal with it, let us discuss the speed considerations that make noise an important factor in digital design.

### How Fast Is Fast?

In this chapter, we will be discussing high-frequency phenomena that actually occur in the frequency range of tens of megahertz. In many interfacing applications, we will be dealing with signals of much lower frequency. For example, the optical shaft encoder discussed in Chapter 2 only produced pulses at a 2500-Hz (hertz). When doing digital design, it is important to realize that even though the principal signals may be slow, the circuits are nonetheless dealing with very high frequencies. Pulses with sharp rise and fall times produce signals whose spectrum contains frequencies related to those rise and fall times and not (as much) to the duration of the pulses themselves. For example, TTL (Transistor–Transistor Logic) circuitry typically has fall times in the vicinity of 2 ns (nanoseconds). The principal frequency component associated with an event 2 ns in duration is 500 MHz! Furthermore, a conventional TTL circuit may be fast enough to detect a noise spike 10 ns in duration, which may be capacitively coupled into a transmission line from some external noise source.

Thus, even though we may believe we are designing a system to deal with very slow signals, and even slow clocks, it is nonetheless necessary to take high-frequency phenomena into account.

Noise is a difficult subject for electrical engineers. Rigorous treatment of noise requires extensive use of both statistics and electromagnetic field theory, areas in which many computer engineering students do not have an extensive background. The alternative is to present a set of "rules of thumb" even though such rules often lack proper justification and, therefore, do not provide the student with insight needed to deal with unexpected situations.

Thus, the topic of noise tends to get lost in electrical engineering curricula, except for a mention of "noise margins" in electronics courses. In this chapter, we will lean toward the rule-of-thumb approach, justifying the rule whenever possible in the context of this book, for noise is a subject that cannot be ignored when interfacing robots to computers, and the rules are adequate, although certainly not complete.

## 3.1 SOURCES OF NOISE

Electromagnetic radiation may enter digital circuits and cause spurious and erroneous signals. Before we can discuss techniques for dealing with noise, we should identify the sources, for different sources require different solutions. Sources of noise are (Morris, 1977)

1. External noise: circuit breakers, commutator brushes, arcing relay contacts, magnetic field–generating activity
2. Power line noise: generated by the same external sources, but coupled to the digital system through the power line
3. Cross-talk: signals coupled from adjacent lines
4. Noise generated in stray impedances from the flow of signal currents
5. Transmission line reflections
6. $I_{cc}$ spikes, from differences in transistor switching speeds

Sources of type 1 are most prevalent in the manufacturing environment. Large motors which are switched suddenly produce large spikes of RF (radio frequency) due to the demand of rapidly changing the current in an inductor. Step motors are particularly difficult to deal with if analog circuitry is present. To see the effects of external noise in a typical robotic installation problem, and to begin learning how to deal with noise, we will consider in Section 3.2 one example, that of noise on encoder signal lines.

## 3.2 EXTERNAL NOISE AND THE SCHMITT TRIGGER

We noted in Chapter 2 that "it is essential that no pulses be permitted whose duration is less than the shortest allowable clock period in the electronic counting logic. This may be easily accomplished by using a low-pass filter. . . ." But the process of low-pass filtering the encoder output does more than simply eliminate the very short, jitter-induced pulses. It also corrupts the rise and fall times of the digital outputs of the encoders. The long rise and fall times produced by filtering can cause improper switching of the gates to which they are input. Furthermore, the encoder output, while it is changing, is essentially an analog signal and is especially susceptible to noise. Even a small amount of RF induced noise on top of such a signal can cause erroneous switching.

The Schmitt trigger circuit provides a simple and effective solution to this and many other similar problems. A diagram of a typical Schmitt trigger circuit is shown in Figure 3.1. This circuit is useful because it has a nonsymmetric voltage transfer characteristic which exhibits different input threshold levels for positive going and negative going signals. We can observe from Figure 3.1 that the Schmitt trigger circuit involves a memory composed of two inverters (transistors $Q_1$ and $Q_2$), which

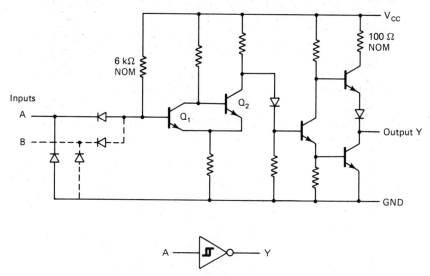

**Figure 3.1**   Schmitt trigger inverter. (*Courtesy of Texas Instruments Incorporated*)

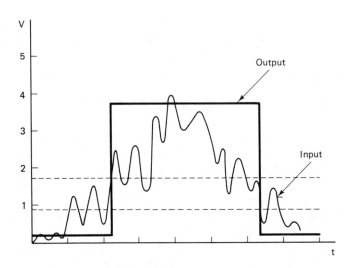

**Figure 3.2**  Input and output from a noninverting Schmitt trigger receiver. The dotted lines represent the switching thresholds.

share a common emitter resistor.  When $Q_2$ is drawing current, the voltage drop across the emitter resistor changes its input switching threshold for $Q_1$.  Hence, it remembers its input "state."  If the device is in the high state, the input must drop to 0.9 V before the device will switch.  Once in the low state, the device remains in that state until the input exceeds 1.7 V.

**Example 3.1  Effects of Thresholds on Schmitt Trigger Performance**

Figure 3.2 shows a noisy pulse which is input to a noninverting Schmitt trigger circuit.  Assuming the thresholds are 0.9 V and 1.7 V, show the output of the circuit.

**Solution:**    See Figure 3.2.

The condition in which the switching threshold depends on the present state of the device is known as *hysteresis*.

The Schmitt trigger's memory enables the device to output properly a single pulse even for the input of Figure 3.2, which a single threshold

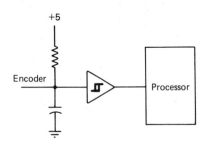

**Figure 3.3**    Using a low-pass filter and a Schmitt trigger receiver to filter out noise on a computer input.

device cannot do. Figure 3.3 shows a circuit utilizing a low-pass filter and Schmitt trigger input inverter to filter and suppress noise on lines between an encoder and a computer input.

## 3.3 CROSS-TALK

Whenever two signal-carrying conductors are placed in proximity, the conductors and the insulator between them constitute a capacitor through which high-frequency signals may travel. Thus, a signal on one line may induce a signal on another, and vice versa.

This particular noise source is most easily dealt with by simply preventing its occurrence. Do not run wires that carry high-speed signals (and that means virtually all digital signal wires) close to each other for long distances. When using flat ribbon cable, make every other conductor ground, or use a cable consisting of twisted pairs, where one of each pair is ground, or use coaxial cable.

On circuit boards, avoid long, closely spaced runs. This is especially critical if one run carries an analog signal and another a digital.

By following these simple and intuitive rules, in combination with the techniques for noise elimination to be discussed in the following sections, it is possible virtually to eliminate cross-talk as a noise source.

## 3.4 SHIELDING: SUPPRESSION OF EXTERNAL NOISE

Every conductor in an electronic circuit is a potential receiving antenna, and every digital gate is a potential RF detector. If a strong source of RF radiation is near enough, these signals can be detected and cause false triggering in digital systems.

Heavy industry, where most industrial robots find application, is replete with such sources. Examples include large electric motors and generators with brushes that arc; relays and switches with arcing contacts; welders; SCRs (silicon-controlled rectifiers), which may switch massive currents on and off extremely rapidly; and many others.

The most effective approach to dealing with such external noise is to suppress it by enclosing the entire digital system in a metal housing. To understand this, recall that no electric field can exist in a perfect conductor. Thus, the electric field component of any electromagnetic wave is blocked and cannot exist inside the shield. If the shield is not a perfect conductor, then the analysis is somewhat more complex: We assume the shield material is a "good" conductor. That is, its conductivity satisfies

$$\sigma \gg \epsilon \omega* \tag{3.1}$$

where $\omega$ is the frequency of the incident wave. Then, the magnitude of the electric field is

$$E_y = E_0\, e^{-x/\delta}\, e^{-j(x/\delta)} \tag{3.2}$$

where

$$\delta = \sqrt{2/\omega\mu\epsilon} \tag{3.3}$$

$x$ is measured as increasing positively into the shield, and $\mu$ is the permeability of the material.

At $x = \delta$,

$$|E_y| = \frac{E_0}{e} \tag{3.4}$$

At this point $E_y$ has decreased to $1/e$ (36 percent) of its initial value. $\delta$ is called the $1/e$ depth of penetration.

**Example 3.2  Effects of Shielding on RF Noise**

For a copper shield, we find that

$$\delta = \frac{6.6 \times 10^{-2}}{\sqrt{f}\ \text{(frequency)}}$$

Assuming that we want to ensure that a 10-MHz wave is attenuated to less than 1 percent of its original value, how thick should the shield be?

**Solution:**

$$\delta = \frac{6.6 \times 10^{-2}}{\sqrt{10^7}} = 20.9 \times 10^{-6}\,\text{m}$$

The wave attenuates to 0.368 of its original value every $\delta$ m. Therefore, since $(0.368)^5 < 0.01$, we provide shielding $5\delta$ thick:

$$5\delta = 5 \times 20.9 \times 10^{-6} = 104.5 \times 10^{-6}\,\text{m}$$

So a shield of well under 1 mm of copper will attenuate a 10 MHz wave to less than 1 percent of its incident value. Higher frequencies will be attenuated even more.

To make use of a shield properly, we must ground it. The correct way to do this is presented in the next section.

---

*$\epsilon$ = permittivity of the material, $\epsilon = \epsilon_r\epsilon_0$, where $\epsilon_0 = 8.85 \times 10^{-12}$ farads/meter, the permittivity of free space, and $\epsilon_r$ is the relative permittivity of the material.

## 3.5 GROUNDING

The preferred way to reduce electromagnetically induced external noise
is shielding, as discussed in the previous section.  Shielding, however,
also has undesirable effects which we must take into account.

Figure 3.4 shows this effect clearly.  The shield is a large sheet of
metal surrounding the circuit.  As such, a capacitance exists between the
shield and the various components of the circuit.  As Figure 3.4 shows,
the shield can connect the input to the output of an amplifier via these
parasitic capacitances.  The capacitance in question may range in value
from a few picofarads up to a few hundred picofarads.  The equivalent
circuit, Figure 3.5, shows clearly the coupling effect of the shield.  The
effect of the shield is thus (in this example) to provide a feedback path
across the amplifier.  This feedback produces the highly undesirable
effect of reducing the high-frequency gain of the amplifier.

In other circuit configurations, the effect of wire-to-shield capacitance may be to provide a signal path from one point to another, where
no path was desired.  In such cases, the shield acts as a source of
cross-talk.

The solution to this problem is simply to ground the shield.  This
converts the equivalent circuit of Figure 3.5 to that of Figure 3.6,
eliminating the feedback.

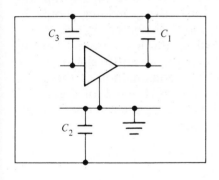

**Figure 3.4**  (Stone) Parasitic capacitances between a typical circuit and shield.

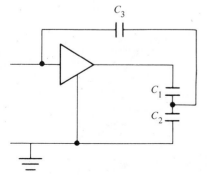

**Figure 3.5**  (Stone) Equivalent circuit indicating
shield to circuit capacitances.

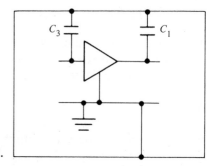

**Figure 3.6**    (Stone) Effect of grounding the shield.

Item 4 in the list of noise sources (Section 3.1) mentions "noise generated by stray impedances throughout the circuit." Ground wires are probably the major source of such problems, for any wire has both inductance and resistance, and when we design assuming that ground is at the same potential everywhere, we make an erroneous and potentially dangerous assumption. In our discussion of grounding, we distinguish between grounding of subsystems within the same chassis and grounding between chassis.

### Grounding Within the Same Chassis

Standard procedure for establishing ground reference within the same chassis follows two guidelines. First, the grounding should be accomplished via a tree of interconnections as shown in Figure 3.7. Second, the shield should be connected to the logic ground with as physically short a wire as possible, so that the circuit-to-shield capacitance can be effectively shorted out. The combination of these two rules requires that care be taken to make legs $A$ and $B$ of Figure 3.7 short. This is most easily accomplished by making the top node a single block of metal, often referred to as a *bus bar*.

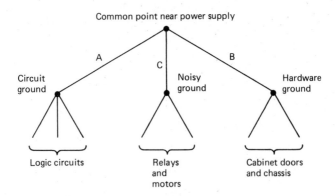

**Figure 3.7**    Grounding tree. (*Courtesy Motorola*)

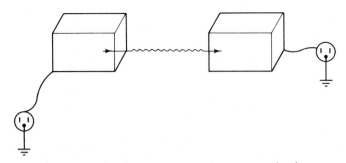

Figure 3.8    Use of earth ground on separate chassis.

**Grounding Between Chassis**

Figure 3.8 represents two chassis with interconnecting signal wires, and with grounds not connected directly to each other, but individually connected to earth ground via the standard three-prong plug. If the two chassis are separated by considerable distance, say, across a plant or in different buildings, ground potential may vary by several volts, making signaling impossible. Worse yet, suppose that it is deemed necessary to provide the interconnecting wire with a shield, as when coaxial cable is used. Now, the low impedance of the shield, connected to ground at both ends, may draw very large currents and even burn.

Variations in ground potential are not uncommon in industry, particularly where large electrical machines are starting and stopping and placing enormous current loads on the grounding system. Figure 3.8 shows a grounding technique that is not only ineffective but dangerous. Two chassis are connected to local earth grounds via the third prongs. If the grounds for the driver and receiving circuit are also connected to their respective chassis, the two ends of the line may vary significantly in their voltage, just because of the grounding difference.

When chassis-to-chassis interconnection to establish ground reference is impossible or infeasible, two options exist: optical isolation and balanced electrical connection.

The circuit diagram of an optical isolator is shown in Figure 3.9. It consists of an LED (light-emitting diode) and a phototransistor. The circuits on either side are electrically isolated. As of this writing, the state of the art in optical data communication is changing so rapidly that it is impossible to discuss price or performance accurately. Such optical techniques are likely to be the standard method for industrial

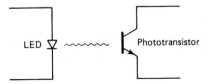

LED    Phototransistor

Figure 3.9    Optical isolator.

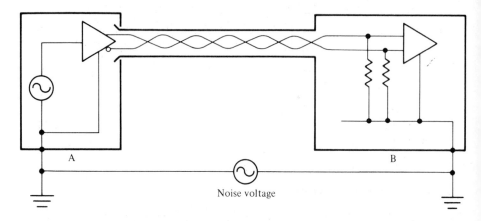

**Figure 3.10**    (Stone) Balanced driver and receiver pair.

communication in the future.    This is especially true of fiber optic
systems.

The other method for communicating between systems at different
ground potentials is the use of balanced drivers and receivers, as shown
in Figure 3.10.    The balanced driver transmits both the signal and its
complement into a twisted pair of wires. The balanced receiver responds
only to the *difference* between the two signals. Furthermore, since both
inputs of the receiver have high impedances, very little current flows
over the wire.    Note that shielded cable is recommended, with the shield
connected to ground *only* at the receiving end.

Balanced drivers and receivers will be discussed in more detail in
the discussion of transmission lines in Section 3.6.


## 3.6 TRANSMISSION LINE EFFECTS

In this section, we will examine the effects of transmission line phenomena and demonstrate how reflections on such lines cause unexpected
electrical effects akin to noise.  First, we will determine under what conditions a wire must be treated as a transmission line.

We say that a conductor is "long" if the amount of time required
for a signal to travel down that conductor is significant relative to the
response time of the attached digital circuitry. To see this more clearly,
we need to consider the speed of modern digital circuits.

In such circuits, propagation delays are typically measured in nanoseconds.  We can compute how far light can travel in one nanosecond
by multiplying by the speed of light.

$$d = 3 \times 10^8 \text{ m/sec} \cdot 10^{-9} \text{ sec}$$
$$= 0.3 \text{ m, or about 1 ft}$$

So light in a vacuum moves at 1 ft per nanosecond. An electrical pulse in a conductor may move considerably slower. In general, the effective velocity of a signal down a transmission line varies as $1/lc$, where $l$ and $c$ are the effective inductance and capacitance of the line, respectively. For most lines of interest, this works out to very close to the speed of light. Therefore, for the rough approximations required at this point, we will take the signal speed to be that of light.

If the round-trip delay for a pulse down a wire is sufficiently long to be detected by the logic family being used, then that wire should be treated as a transmission line.

For example, suppose we have a line 5 ft long driving Schottky TTL logic. Such logic is capable of detecting pulses as short as 4 ns. A reflected pulse down the 5-ft line will return to the source 10 ns after it was placed on the line, potentially interacting with the original pulse and causing spurious signals by its interaction. Thus, a 5-ft wire driving Schottky TTL must be treated as a transmission line.

For digital systems of clock rates under 10 MHz, this typically means that transmission line effects may be ignored within a circuit board but should be considered for the interconnections between boards.

When a wave or pulse is propagated down a transmission line, it is reflected from the terminating end of the line, producing a *reflected wave*. The magnitude of the reflected wave may be related to the voltage impressed on the line by

$$\frac{V_R}{V_0} = \frac{Z_1 - Z_0}{Z_1 + Z_0} \qquad (3.5$$

The ratio $V_R/V_0$ is known as the *reflection coefficient*, and it depends on $Z_0$, the characteristic impedance of the line, and on $Z_1$, the load (termination) impedance.

When $Z_1 = Z_0$, we say that the termination impedance is matched to the line, and no reflection occurs. While this condition is clearly desirable, it is not always easy (or even possible) to achieve, as we will demonstrate. However, first, let us examine the effects of unmatched line termination on signals.

### Example 3.3  High-Impedance Termination

A transmission line is unterminated. That is, at the far end of the line (most distant from the observer), the line ends in an open circuit. Assuming that a voltage step of 1 V is introduced on the line at $t = 0$ and assuming that waves require 5 ns to travel down the line, graph the voltage at both ends of the line as a function of time.

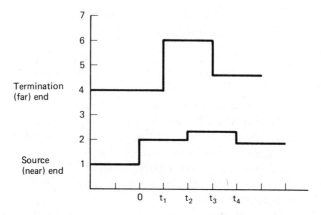

**Figure 3.11(a)**   Voltage at near (transmitting) and far (receiving) ends of a transmission line, assuming that a perfect voltage step is placed on the line at $t = t_0$.

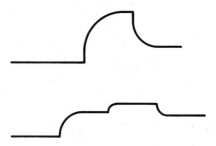

**Figure 3.11(b)**   A more realistic representation of the response of the line in Figure 3.11(a), assuming nonzero rise times.

**Solution:**   In Figure 3.11(a), we show the voltage at both ends of a line, assuming that a voltage step is introduced at time $t = 0$. If $Z_1 > Z_0$, we see that the reflected wave is of the same sign as the incident wave. In the limiting case (open circuit or unterminated line), the reflection coefficient approaches +1 and the reflected wave is exactly equal to the incident wave. At the far end, we observe a voltage which is the sum (to a reasonable approximation) of the two waves. Point $t_1$, in Figure 3.11(a) shows this moment.

Then, sometime later, the reflected wave reaches the source. Let us make the assumption that the source impedance is low, considerably lower than the characteristic impedance of the line.* If $Z_s < Z_0$, then the reflection coefficient seen by the returning reflected wave is nearly − 1, for purposes of illustration, let us say, it is − 0.7. At this point [$t_2$ in Figure 3.11 (a)], the voltage at the source is the sum of the original voltage step (1 V), the reflected wave (1 V), and the second direct wave (− 0.7 V). The first reflected wave and the second direct wave nearly cancel each other and result in a small increase in voltage at the source. At time $t_3$, the second

*This assumption is reasonable, since the lower the source impedance, the greater its drive capacity. A high-impedance source could not supply enough current to drive the loads.

direct wave arrives at the far end and is reflected without inversion, reducing the end voltage from 2 V to 0.6 V. This second reflected wave arrives at the source at $t_4$ and is inverted again. This process of oscillation continues, until stability is reached some time later.

In realistic lines, propagation delays are on the same order as rise time; therefore, the square pulses of Figure 3.11(a) do not occur, and the signals are more likely to look like Figure 3.11(b).

Figure 3.12 shows another way to look at waves as they propagate down a transmission line. In this figure, the diagonal lines indicate each wave. The time of arrival of each wave is explicitly shown, as is the voltage on the line at that point. Remember, *the voltage is the sum of all the waves*. For example, at time $t_3$, the voltage at the end of the line is the sum of the first and second direct waves and the first and second reflected waves.

### Line Matched at the Source

Figure 3.13 illustrates a line terminated at the source in an impedance which matches the line characteristic impedance. In this case, the magnitude of the first direct wave is equal to the source voltage

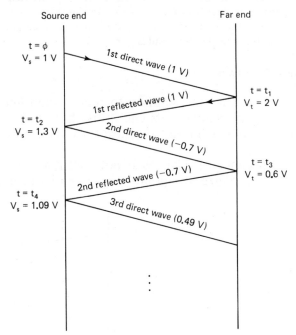

**Figure 3.12** Bounce diagram for the wave sequence shown in Figure 3.11. The instantaneous time (e.g., $t_3$) is defined to be an instant *after* the wave arrives; hence, the voltage of the reflected wave must be included.

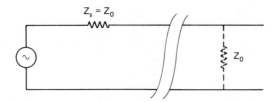

Figure 3.13   Source termination.

divided between the two impedances (the matching resistor and the line). The following example shows the effect of source termination.

**Example 3.4  Line Matched at the Source**

A transmission line of characteristic impedance $Z_0$ = 100 $\Omega$ (ohms) is terminated at the source in 100 $\Omega$ and at the far end in 300 $\Omega$. Construct a bounce diagram showing the waves resulting from a 10-V step output from the source at $t = 0$.

**Solution:**   At $t = 0$, the 10-V signal from the source sees two impedances, the 100-$\Omega$ source impedance in series with the 100 $\Omega$ line impedance. The 10 V divides equally between these, resulting in a first direct wave of 5 V, as shown in Figure 3.14.

Since the source impedance is 0 V, there is no second direct wave. The voltage at both ends of the line then assumes a constant 7.5 V. Note that this is in agreement with a steady-state analysis of the circuit. In the steady state, the line has 0 impedance, and the 10 V from the source is divided between the 100 $\Omega$ source and 300 $\Omega$ termination impedances, resulting in 7.5 V.

**Choosing a Terminator**

It would seem that the solution to transmission line problems is simple: terminate every line at its end in its characteristic impedance. However, such an approach is not practical for several reasons:

1. The impedance of all lines is not well defined. The characteristic impedance of a point-to-point wire, run above a ground

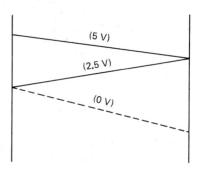

Figure 3.14   Wave on a line terminated at the source.

plane, for example, has an impedance that can vary with the physical separation between the conductor and ground.

2. It is not reasonable or necessary to terminate every wire. For short runs and/or slow signals, it makes perfect sense simply to wait for the ringing to die out.

3. Terminators consume power. As shown, the source termination divided the source voltage in half initially, with the line slowly coming up to the source voltage as the transients die out. An end terminator consumes power in the steady state.

As we shall see in the next section, in many situations, the termination problem is even more complex.

**Buses**

Figure 3.15 illustrates a transmission line terminated in its characteristic impedance at both source and load but which is tapped by a second line, also terminated. The second (presumed shorter) line is referred to as a stub.

Initially, the drive voltage is split evenly, producing a first direct wave of $V_0/2$. At the point where the stub meets the line, the first direct wave is confronted with two impedances of $Z_0$ in parallel, resulting in a reflection coefficient of

$$\frac{z_0/2 - z_0}{z_0/2 + z_0} = -\frac{1}{3}$$

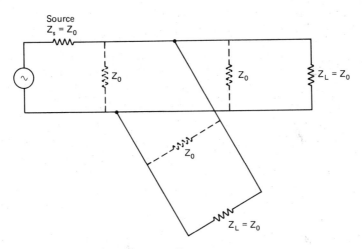

Figure 3.15    A terminated line with a stub.

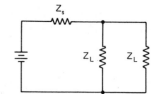

**Figure 3.16** DC equivalent circuit of
the line in Figure 3.15.

So $\frac{1}{3}$ of the direct wave is reflected, resulting in a reflected wave of
$V_r = -\frac{1}{3} (V_0/2) = -V_0/6$, and the voltage at the tap immediately drops to

$$\frac{V_0}{2} - \frac{V_0}{6} = \frac{V_0}{3}$$

This voltage wave propagates down both the bus and the stub. However, the current splits evenly. Since no further reflections occur, this
should be the steady state. Analysis of Figure 3.16 in the DC case
shows this to be correct.

In this example, we have shown a system with matched termina-
tions everywhere which still possesses reflections from the taps, and all
the steady-state power consumption and driver loading problems still
exist. Thus, terminating each stub as is done in this example is neither
practical nor effective.

A solution to this problem is to use stubs that are not transmission
lines. Such an approach is not as naïve as it may sound, for by using
care in the physical layout of boards as they plug into a backplane
bus, and using high-impedance receivers, the effects of taps can be minimized. Again, the *physical* layout is critical.

### Active Termination

Rather than using a single resistor between the line and ground, a
line may be terminated by a resistor connected to a power supply, as
shown in Figure 3.17. Since the power supply acts as a short circuit to
the high frequencies of wave fronts, the use of a resistor equal to $Z_0$ will
properly terminate the line. Furthermore, by setting the power supply

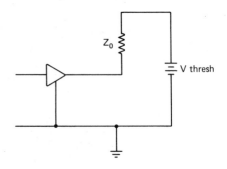

**Figure 3.17**   Active termination.

halfway between signal HIGH and LOW levels, power requirements from drivers are reduced, and noise immunity is improved.

To see this, consider an unterminated bus driven by a tristate device. When the driver is disabled, the line gradually discharges. When enabled, the driver must then charge the line to as much as $V$. With active termination, the unused line floats at $V/2$, and the driver is only required to change the potential by an additional $\pm V/2$. This doubles the charging speed of the line.

This type of active termination will not work for open collector buses. Open collector devices have no idle state and require a pull up resistor between the bus and $V_{cc}$.

### Bus Interconnections

After active termination, source termination (Figure 3.13) is the most popular technique for dealing with reflections. However, since a bus may have more than one source, this technique is not completely effective either.

Even with active termination, some reflections on buses are unavoidable, due to reflections at the taps. Below 10 MHz, the procedures outlined earlier—using short stubs with high impedance receivers and keeping the bus short (less than 0.3 to 0.5 m)—will be adequate. Of course, some termination should be used at the end of the longest bus.

If it is necessary to use a longer bus, say, between chassis, the bus should be divided into portions that are short and dense, with longer, untapped cables between portions. Use of a single driver and receiver on these longer cables will allow the use of precise termination techniques.

### Cable Connections

Between boards, connections may be made by use of a printed circuit backplane (a *mother board*), a wired backplane, or cables. If the cable length is quite short, less than about 0.3 m, no special care need be paid to transmission line effects for signals below 10 MHz. For longer cables or higher speeds, termination and shielding need to be considered.

### Twisted Pairs

Probably the most effective technique for cable connections between chassis is the use of multiple twisted pairs of wires, and differential (or "balanced") line drivers and receivers, which were discussed in Section 3.5. Cables composed of many twisted pairs are widely available, both as round, jacketed cable, and in flat ribbon form.

**TABLE 3.1    Outputs of a Differential Line Driver**

| Input | $V +$ Output | $V -$ Output |
|-------|--------------|--------------|
| LOW | $V_{ref}$ | $V_{ref}$ |
| HIGH | $V_{ref} + V_0$ | $V_{ref} - V_0$ |

A differential driver has two outputs, one positive (with respect to an arbitrary reference point) and one negative. Table 3.1 shows the outputs of a noninverting differential driver for inputs of logic HIGH and LOW.

The differential receiver, in turn, responds to the difference of its input, and has an input switching threshold which is greater than either input alone.

The key idea in the use of differential drivers and receivers is that external noise due to sources 1, 2, or 3 will cancel itself. To see this, consider a twisted pair carrying a signal $s(t)$ on one conductor and $-s(t)$ on the other. Now, suppose that a nearby source of electromagnetic radiation couples a noise voltage $n(t)$ onto both wires. One conductor now carries $s(t) + n(t)$, and the other carries $-s(t) + n(t)$. At the receiver, the two inputs are (effectively) subtracted,

$$s_r(t) = [s(t) + n(t)] - [-s(t) + n(t)]$$

$$= 2s(t)$$

and the noise voltage has canceled itself.

Of course, this additive model of noise is only a first approximation, and sufficiently large noise sources can induce false triggering even on differential pairs. Use of such balanced drivers and receivers is, however, accepted industrial practice, nontemporal as it is the most effective means currently available for dealing with external noise coupled into long cables.

This technique is sometimes referred to as *double-ended signaling.* ECL (emitter-coupled logic) systems use double-ended signaling as standard procedure between almost all circuits, both because of the very high speeds of ECL and because the ECL technology inherently uses differential receivers.

### Terminating Balanced Lines

Figure 3.18 shows four different techniques for terminating a balanced line. Figures 3.18(a) and (b) are to be preferred over (c) and (d), since (c) and (d) require that the driver (in the steady state) supply current to a small resistance returned to ground.

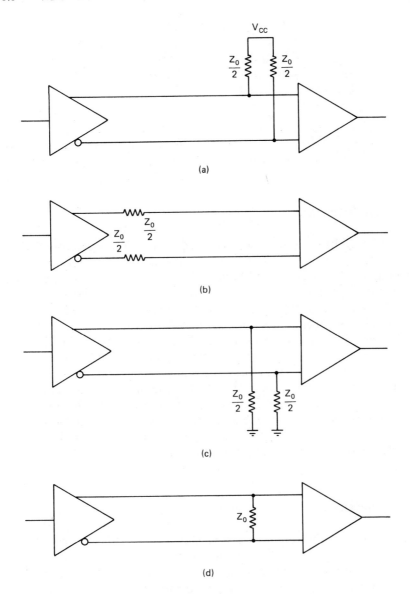

**Figure 3.18**    (Stone) Terminating balanced lines.

The source termination technique of Figure 3.18(b) requires less power than does the load termination technique of Figure 18(a) and is, therefore, preferred. (Remember, higher power consumption means higher ground currents, which increase the related noise problems discussed in Section 3.4). However, source termination requires that receivers wait longer to be sure that the line is stable.

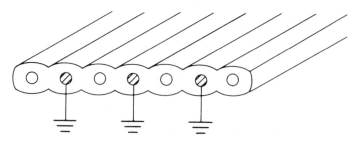

Figure 3.19    (Stone) Use of ground in flat ribbon cable.

### Flat Ribbon Cable

Flat ribbon cable is widely used today for board-to-board inter-connection, and even for chassis-to-chassis interconnection over short distances. Although flat ribbon cable made from twisted pairs is available, more common and inexpensive cables contain 20 to 50 conductors laid out in parallel runs. Such cables are especially sensitive to cross-talk and ground return problems. Fortunately, the same simple technique helps to solve both problems. As is shown in Figure 3.19, providing a ground wire between every pair of signal conductors effectively isolates the conductors and eliminates cross-talk from adjacent conductors. Furthermore, this technique provides adequate ground return for whatever signal currents are carried over the cable. It is in general *not* adequate to serve alone as a chassis-to-chassis ground.

The disadvantage of interleaving signal and ground wires is, of course, that it doubles the amount of cable required. Nonetheless, it is recommended for any cables over 1 m long.

### 3.7  $I_{cc}$  SPIKES

The term $I_{cc}$ *spike* refers to a very short pulse of high current flowing in the collector circuit (and hence in the power supply wires) of TTL circuits. Although similar phenomena occur in other technologies, it is most pronounced in TTL and will be discussed in that context.

$I_{cc}$ spikes are transient phenomena resulting from the fact that transistors require more time to come out of saturation than to go into saturation. This is because going into saturation is an active process—injecting charge into the base—whereas coming out of saturation is a passive process—drawing charge out of the base.

Figure 3.20 shows the totem pole output of a TTL device at just the critical moment, when $Q_1$ has just been switched "on," and $Q_2$ has not yet come out of saturation. At this instant, we find that the emitter

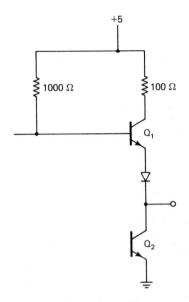

**Figure 3.20**   TTL totem pole output.

of $Q_1$ is at 0.9 V [assume $V_{be}$ (on) = 0.7 V, $V_{ce}$(sat) = 0.2 V] and that the base of $Q_1$ is at 1.6 V. Thus, $Q_1$ has a base current of (5 - 1.6)/1000 = 3.4 mA. Assuming any sort of reasonable beta for $Q_1$ will lead us to the conclusion that $Q_1$ is saturated. It should be emphasized that this is a transient phenomenon, and $Q_1$ is never saturated in the steady state. Thus, the voltage across the 100 Ω pull-up resistor is

$$V_r = V_{cc} - 2V_{ce}(sat) - V_d$$
$$= 5 - 2(0.2) - 0.7$$
$$= 3.9 \text{ V}$$

So $I_c$ = 3.9/100 = 39 mA.

Suppose that we have a 32-bit register whose output is simultaneously switched from 00000000 to FFFFFFFF. For a period of about 1 ns, this register demands

$$32 \times 0.039 = 1.25 \text{ A}$$

Can the power supply provide this current? Probably. Can it provide it quickly enough? To answer that question, we need to take a close look at the wire between the 32-bit register and the power supply. Figure 3.21(a) shows the model generally used to represent such wires. Unfortunately, the model of Figure 3.21(b) more accurately represents the wire.

We can see from this model that the power supply may not be able to provide such large currents so quickly, due to the inductance of the power bus. If the power supply cannot provide this level of current

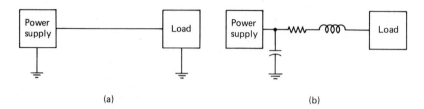

(a)                                                    (b)

**Figure 3.21(a)**  Common model of power    **Figure 3.21(b)**   Realistic model of power
supply wires.                                 supply wires.

rapidly enough, then $V_{cc}$ will drop at the chip. A change in power supply logic voltage is likely to cause false triggering or change of state of a memory device.

A simple solution to this problem is to provide a capacitor between $V_{cc}$ and ground physically very close to each chip. But one final choice must be made—namely, what capacitor to use. Figure 3.22(a) and (b) shows two different models for a capacitor. The inductance of the capacitor is crucial, for a capacitor with high internal inductance is worthless when it is necessary to provide charge at such very high speeds.

Two types of capacitors dominate today's digital technology: tantalum and ceramic. These terms refer to the dielectric material. *Tantalum capacitors* provide quite high capacitance for very small physical size. For that reason, they are attractive for use in high-density digital circuits. Unfortunately, tantalum capacitors also have relatively high inductance and are, therefore, ineffective in filtering $I_{cc}$ spikes. *Ceramic capacitors,* on the other hand, work effectively if properly laid out over the circuit. For purposes of this discussion, *monolithic* capacitors may be considered as a type of ceramic. They provide the high speed of ceramic with greatly reduced size.

### Example 3.5  Current Loading

Assume that the 32-bit register just described may be modeled (to a *very rough* approximation) by a 4 $\Omega$ resistor in series with an ideal switch to ground as shown

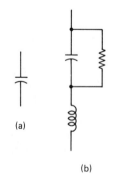

(a)

**Figure 3.22**  Two models of a capacitor, ideal and realistic.                                    (b)

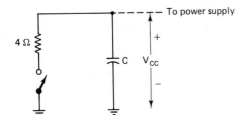

**Figure 3.23** Model of a register during an $I_{cc}$ spike. The switch remains closed for 1 ns. $V_{cc}$ is initially 5 V.

in Figure 3.23. Furthermore, assume that the inductance of the wire from the register to the power supply is so large that no current can flow during the 1-ns $I_{cc}$ spike. Determine the minimum bypass capacitance required to guarantee that $V_{cc}$ never drops below 4.5 V during an $I_{cc}$ spike of 1-ns duration.

**Solution:**  We know that a capacitor initially charged to 5 V decays through a resistor, following

$$V_{cc} = 5\,e^{-t/RC}$$

We solve this equation for $RC$, given that

$$t = 10^{-9} \quad \text{and} \quad V_{cc} = 4.5 \text{ V}$$

$$\frac{4.5}{5} = e^{-10^{-9}/RC}$$

Taking a logarithm, we find $RC = 10^{-9}/0.1054$; therefore,

$$C = 2.37 \times 10^{-10} \text{ F} \quad \text{or} \quad 237 \text{ pF}$$

The usual value of 0.01 $\mu$F is, therefore, more than adequate.

**Circuit Board Layout**

Some recommendations and guidelines concerning the layout of circuit boards are included here, since proper layout is essential to proper operation of high-speed logic.

1. If possible, use a multilayer "sandwich" printed circuit board with a solid ground sheet in the center layer. This provides a degree of shielding and a good, low resistance to ground for signal returns.
2. Bypass the board with a large tantalum capacitor, typically 10 to 20 $\mu$F, for suppression of low-frequency external noise (Figure 3.24).
3. Bypass chips with 0.001 to 0.01 $\mu$F ceramic or monolithic capacitors between $V_{cc}$ and ground located physically as close to the chip as possible. With less complex chips (SSI), it is not necessary to bypass every chip; every two or three is adequate. However, bypassing each chip is good practice, and capacitors are quite inexpensive.

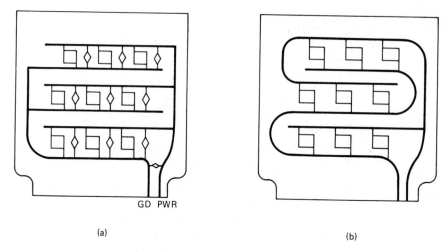

GD  PWR

(a)                                                                                   (b)

**Figure 3.24**    Proper (a) and improper (b) circuit board layouts.

**4.** Avoid long, serpentine layouts of $V_{cc}$ and ground. A tree structure such as that illustrated in Figure 3.24(a) is far better than the serpentine layout illustrated in Figure 3.24(b).

## 3.8 SYNOPSIS

### Vocabulary

You should know the definition and application of the following terms:

active termination
balanced driver and receiver
bounce diagram
bus bar
bypassing
ceramic capacitor
characteristic impedance
cross-talk
external noise
grounding tree
hysteresis
$I_{cc}$ spike
mother board

optical isolator
parasitic capacitance
power line noise
reflection
reflection coefficient
Schmitt trigger
shielding
stub
source termination
spike
tantalum capacitor
transmission line
twisted pair

**Notation**

| Symbol | Meaning |
| --- | --- |
| $\sigma$ | Electrical conductivity |
| $\epsilon$ | Permittivity of a material |
| $\epsilon_0 = 8.85 \times 10^{-12}$ F/m | Permittivity of free space |
| $\epsilon_r$ | Relative permittivity |
| $\omega$ | Frequency, measured in rad/sec |
| $E_y$ | Magnitude of an electric field (in the $Y$ direction) |
| $\mu$ | Permeability: For nonferrous materials, $\mu = \mu_0 = 4\pi \times 10^{-7}$ H/m |
| $f$ | Frequency, measured in hertz |
| $V_R$ | Magnitude of a reflected wave |
| $V_0$ | Magnitude of an incident wave |
| $Z_1$ | Impedance of a terminator |
| $Z_0$ | Characteristic impedance of a line |
| $Z_s$ | Output impedance of a signal source or impedance inserted at the source end of a line |

## 3.9 REFERENCES

BLOOD, W. R. *MECL system design handbook.* Mesa, Ariz.: Motorola Semiconductor Products, 1971.

MORRISON, R. *Grounding and shielding techniques in instrumentation.* New York: Wiley-Interscience, 1977.

MORRIS, R., & MILLER, J., eds. *Designing with TTL integrated circuits,* Texas Instruments Electronics Series. New York: McGraw-Hill, 1971.

STONE, H. *Microcomputer interfacing.* Reading, Mass.: Addison-Wesley, 1982.

Texas Instruments. *The TTL data book.* Dallas, Tex.: Texas Instruments, Inc.

## 3.10 PROBLEMS

1. A particular TTL circuit has a 1-ns long $I_{cc}$ spike. During that time, the circuit demands 1.25 A. Assume the inductance of the conductor to the power supply is so large that no current can come from the supply and that all must be provided by the bypass capacitor. What is the minimum size of bypass capacitor required to guarantee that by the end of the spike, $V_{cc}$ to the circuit has not dropped below 4 V? *Hint:* To a first approximation, treat the circuit as a resistor. Find its resistance, and compute time constants.

2. A copper wire 50 cm long feeds a Schottky TTL gate. Assume that the speed of an electrical signal on the wire is 0.6 times the speed of light. The Schottky gate will respond to any pulse over 2 ns in length. Is it necessary to treat this wire as a transmission line?

3. Repeat problem 2 for a 74LS series gate that responds to any pulse over 6 ns long.

4. Assume that the wire in problem 2 is laid along a ground plane so that it acts as a transmission line with 100 $\Omega$ of characteristic impedance. Assume the transmitting gate has 10 $\Omega$ of output impedance and the receiving gate has infinite input impedance. Sketch the voltage at the source end of the line as a function of time for a 10-V step asserted by the source at $t = 0$.

5. Repeat problem 4 with source termination matched to the line.

6. Figure 3.P.1 shows a signal input to two inverters. INV1 is a conventional TTL inverter, 7404. INV2 is a Schmitt trigger inverter, 7414. Use the specification sheet for the 7414. Assume that the 7404 acts as a perfect switch, with threshold at 0.8 V. Sketch the two output voltages, $V_{01}$ and $V_{02}$.

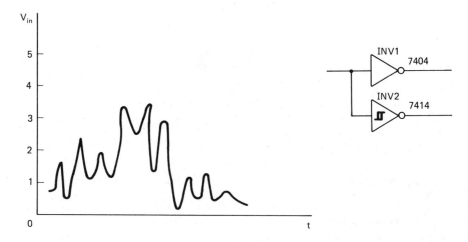

Figure 3.P.1

# 4

## *ACTUATORS*

Actuators are the devices that make robots move. There is a plethora of such devices in the field, and we will restrict our attention in this chapter to four: step motors, DC motors, hydraulic pistons, and pneumatic pistons. The emphasis will be on DC motors and hydraulic systems because of the prevalence of these types of actuators in modern industrial robots. In studying the similarities and differences between these two types of devices, we will learn fundamental principles which we will be able to apply to control of many types of actuators.

### 4.1 THE DC MOTOR

As do all electromechanical devices, the DC motor makes use of the fact that a wire carrying a current in a magnetic field experiences a force. In a DC motor, the windings wrapped around a rotating armature carry the current. An arrangement of commutator segments and brushes (Figure 4.1) ensures that the DC current is always in the same direction relative to the magnetic field, thus resulting in a constant force direction (or torque).

The principal variation among different types of DC motors lies in the mechanism used to develop the magnetic field. In a permanent magnet DC motor, the field is developed, as the name suggests, by permanent magnets. In such a motor, the torque, $T_m$ , is related to mag-

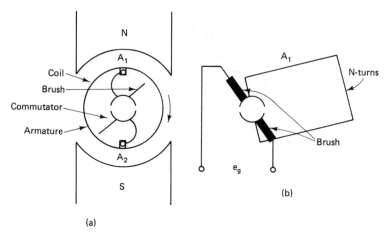

(a)

**Figure 4.1** An elementary DC machine. (a) A cross-sectional view of poles and armature. (b) A view of the armature coil and the slip rings.

netic flux, $\Phi$, and armature current, $I_a$ by

$$T_m = K_t \Phi I_a \tag{4.1}$$

under steady state conditions, where $K_t$ is a proportionality constant. Since $\Phi$ is a constant in a permanent magnet motor, we can say that in the steady state, torque is proportional to armature current.

The magnetic field can also be generated by an electromagnet. If the current for the electromagnet is provided on a pair of wires separate from the armature current, then

$$T_m = K_t K_f I_f I_a. \tag{4.2}$$

where $I_f$ is the current in the field windings and $K_f$ is a constant depending on the number of winding turns and the permeability of the iron around which the windings are turned.

If $I_f$ is supplied by a separate source (Figure 4.2), then the basic

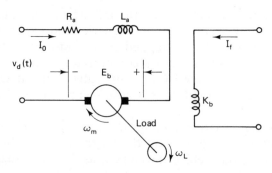

**Figure 4.2** (Gourishankar) An armature-controlled DC motor. (*Courtesy Harper & Row*)

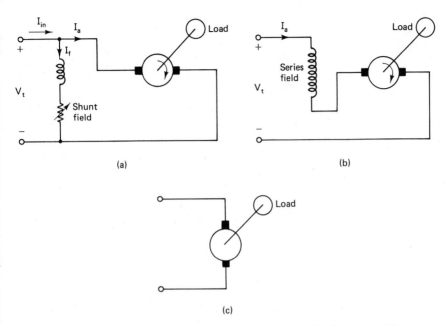

**Figure 4.3**    (Gourishankar) (a) A DC shunt motor; (b) a DC series motor. (*Courtesy Harper & Row*)

torque-current relation of Eq. 4.1 is still true.  In many instances, however, it is desirable to derive the field current from the same pair of wires that provides the armature current.  Figure 4.3 shows two of the more popular ways in which to accomplish this, shunt and series motors. The steady-state equivalent circuits are shown in Figure 4.4.  In these cases, analysis is complicated by the fact that the motor, when turning, acts as a generator, producing the back EMF (electromotive force), $E_b$.

Now, the analysis depends on the driving source.  For example, in the shunt motor, if $V_t$ (Figure 4.4) is held constant, then torque varies linearly with armature current.

In the case of the DC series motor, we have $\Phi = K_f I_a$ and

$$T_m = K_t K_f I_a^2 \qquad (4.3)$$

and the torque varies as the square of the current.

Most "servo" motors are permanent magnet motors and so obey

$$T_m = K_{pm} I_a \qquad (4.4)$$

where $K_{pm}$ results from the product of $K_t$ and $\Phi$ in Eq. 4.1, with $\Phi$ a constant.  This is the model that we will use to analyze the behavior of servo systems in Chapter 5.  The other types of DC motors, which drive their field magnetization electrically, may likewise be analyzed, but the

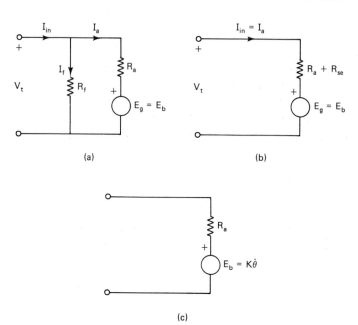

(a)                    (b)

(c)

**Figure 4.4**  (Gourishankar)  Equivalent  circuits  of  the  motors  in  Figure  4.3.
(*Courtesy Harper & Row*)

analysis is more complex than that required for presentation of fundamental robotics concepts.

### 4.1.1  Loading DC Motors

The primary loads on motors are friction, inertia, and constant or varying torque loads.  At this point, we will consider only friction and inertia, designated $F$ and $J$, respectively.  A rotating physical system, in the absence of outside forces, obeys

$$T = J\ddot{\theta} + F\dot{\theta} \qquad (4.5)$$

where $T$ is the torque exerted, $\theta$ is the angular position measured in radians, $\dot{\theta}$ is angular velocity in radians per second, and $\ddot{\theta}$ is angular acceleration measured in radians per second squared.  The frictionless version of this relation,

$$T = J\ddot{\theta} \qquad (4.6)$$

may be seen to be the rotational dual of Newton's law,

$$F = ma \qquad (4.7)$$

Joint

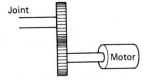

**Figure 4.5**  $N$ is the number of teeth on the larger gear divided by the number of teeth on the smaller gear.

DC motors typically are capable of high rotational velocities and relatively low torque. Therefore, gearing is used to trade decreased speed for increased torque (Figure 4.5.). If $N$ represents the gear ratio, we find

$$T_{\text{(applied to load)}} = N \cdot T_{\text{(applied by motor)}} \qquad (4.8)$$

$$\theta_{\text{load}} = \frac{1}{N}\, \dot{\theta}_{\text{motor}} \qquad (4.9)$$

The load is divided by the square of the gear ratio, resulting in equations for the equivalent inertia and friction seen by the motor, $J_{eq}$, and $F_{eq}$, respectively.

$$J_{eq} = J_a + \frac{1}{N^2}\, J_1 \qquad (4.10)$$

and

$$F_{eq} = F_a + \frac{1}{N^2}\, F_1 \qquad (4.11)$$

where $J_a$ and $F_a$ are the inertia and friction of the motor itself and $J_1$ and $F_1$ are the inertia and friction of the load. Gear ratios of $100:1$ are not uncommon, and Eq. 4.10 says that in this case, the inertia of the load is divided by $10^4$ before being perceived by the motor. Thus, while the armature inertia of the motor itself may seem at first to be trivially small compared with the load inertia, the gearing may make the exact opposite true.

Equation 4.11 should be taken with a grain of salt, since it assumes frictionless gears. In robot systems, most of the friction is likely to be in the gears themselves, and Eq. 4.11 does not accurately model this effect. Furthermore, the gear friction varies considerably with the temperature of the lubricant. We will see in Chapters 5 and 11 some techniques for dealing with such nonlinear loads.

**Example 4.1  Loading of a DC Motor**

A permanent magnet DC motor with armature inertia of

$$5 \times 10^{-3} \text{ kg-m}^2$$

is used to drive a load of mass equal to 10 kg. The load is located at an (effective) radius of 0.5 m. A 100:1 gearing is used. Find the equation relating torque to angular rotation. Ignore gear inertia and friction.

**Solution:**    The load inertia is divided by the gear ratio squared, producing

$$J_{leff} = \frac{I}{N^2} = \frac{M_1 r^2}{N^2} = \frac{(10)(0.5)(0.5)}{(100)^2} = \frac{2.5}{10^4}$$

$$J_{leff} = 2.5 \times 10^{-4} \text{ kg m}^2$$

$$J_{eq} = J_a + J_{leff} = 50 \times 10^{-4} + 2.5 \times 10^{-4} = 5.25 \times 10^{-3} \text{ kg m}^2$$

Then

$$T = J_{eq} \, \ddot{\theta}$$

### 4.1.2  Driving the DC Motor

The DC motor provides a torque which, by Eq. 4.1, is directly proportional to the armature current.  (We will henceforth discuss only the permanent magnet DC motor.) Since it will provide this torque, even when stalled, it is desirable to drive the motor with a controllable DC current source.  Furthermore, since it will be necessary to drive the motor in either direction, such a DC source must be capable of supplying both positive and negative currents.  The most straightforward mechanism for accomplishing this is the DC-coupled push-pull amplifier.

#### Push-pull Amplifiers

The push-pull amplifier of Figure 4.6 provides a voltage out at high current levels in response to a signal voltage in.  Such amplifiers are the standard circuit configuration used in *servo amplifiers*, which may be obtained commercially.

Since torque in a DC motor is proportional to current, and we really wish to control torque, it would be very desirable to have an amplifier in which the current out (as opposed to the voltage out) is proportional to the voltage in.  Since a DC motor in motion acts as a generator,* the current through the armature windings is not simply related to the applied voltage.  (This will be discussed in more detail in Section 5.5).  Thus, controlling the voltage, as in the case of the push-pull amplifier, does not directly control the torque.

---

*That is, a voltage, referred to as the "back EMF," is generated that is proportional to motor velocity and is opposing the applied current flow. See Figure 4.8.

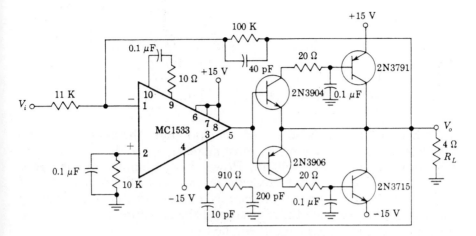

**Figure 4.6**    Push-pull amplifier. (*Courtesy Motorola*)

Adding a sense resistor and an operational amplifier to a push-pull circuit can provide a current out in response to a voltage in. With this simple addition (Figure 4.7), direct control of torque is possible. In this circuit, the motor current is sensed by a resistor that is small with respect to motor resistance, and the resulting signal is fed back to the operational amplifier. Increasing the voltage out of the D/A will cause the current out of the amplifier to increase proportionally.

It should be noted that a control requiring both high current and high speed simultaneously may place unreasonable demands on a power amplifier. To see this, consider the equivalent circuit of a DC motor as shown in Figure 4.8. As the motor turns rapidly ($\dot{\theta}$ is large), a signif-

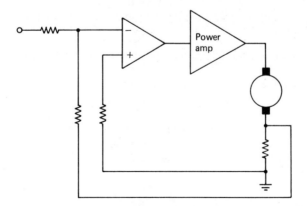

**Figure 4.7**    Addition of a sense resistor and another operational amplifier to a power amplifier provides a controllable current source.

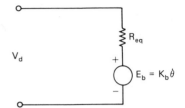

**Figure 4.8**    Equivalent circuit of a DC
motor.

icant back EMF, $E_b$, is generated which opposes the applied $V_d$. To
maintain a constant current, $V_d$ must increase, which the feedback current source of Figure 4.7 will do. However, $V_d$ is limited by the power
supply values, which may be quickly reached at high motor speeds. If
the velocity is so high that $V_d$ cannot exceed $E_b$ sufficiently to maintain
the current flow through the resistance of the armature, the current will
drop and current/torque control will be lost.

A second consideration in designing a motor driver is the fact that
excessive armature current will demagnetize the PM (permanent magnet)
field magnets. This circumstance generally leads to a requirement for
current control.

These observations should lead the reader to realize that simple
linear models of power supplies and actuators are in general not adequate
to predict accurately the performance of robot systems, and we must
strive to develop control schemes that are *robust*, that is, schemes that
will give good performance in spite of our inability to develop a simple
mathematical model of the system.

**Example 4.2    Electrical Load Presented by DC Motor**

The motor described in Example 4.1 is subjected to some tests: When stalled, it
draws 10 A from a 24-V supply. When allowed to run unloaded, it accelerates and
eventually reaches a maximum rotational speed of 3000 rpm (revolutions per
minute). (This is the rotary speed of the *motor* shaft, not the output of the gear,
which turns 100 times more slowly.) Determine the electrical characteristics of
the motor.

**Solution:**    At stall, we determine the resistance of the motor windings in 24/10 =
2.4 Ω (ohms).

Using Figure 4.8, we write a loop equation and determine

$$24 \text{ V} = I(2.4 \text{ } \Omega) + K_b \text{ (3000 rpm)}$$

We appear to have an insoluble problem, since both back EMF gain and current are
unknown. However, we were given one more piece of information. 3000 rpm is
the *maximum* speed attained. Since the speed has "topped out," there is no acceleration, but since

$$T = J\ddot{\theta},$$

no acceleration means no torque. Since

$$T = K_m I$$

no torque means no current, and the loop equation simplifies to

$$K_b = \frac{24}{3000} = 8 \times 10^{-3} \text{ V/rpm}$$

### Pulse-width Modulation

Rather than drive the DC motor with a constant voltage or current, as was described in the previous section, one could drive the motor with a rapidly changing current. If the rate of change of that current is far higher than the response speed of the motor and attached physical system, the net effect will be a response to the DC component of the drive signal.

One popular way in which to achieve this type of control is to use pulses, as shown in Figure 4.9. The repetition rate, $T_c$, is a constant and the pulse "on" voltage, indicated positive in Figure 4.8, will be negative for reverse drive. The ratio $T_+/T_c$ defines the *duty cycle*. The average (DC) value of the drive signal is

$$V_{AV} = V_+ \left(\frac{T_+}{T_c}\right) \tag{4.12}$$

The principal advantage of PWM (pulse-width modulation) over linear control is the simplicity of the drive electronics and the ease of computer interfacing. Figure 4.10 shows a simple circuit adequate for driving a small DC motor using PWM. The computer may be used to control the state of the $A$ and $B$ control lines. Timing may be accomplished via software timing loops or by interrupts from an internal timer. Many microprocessors contain software-settable interval timers that make such functions readily available. Since the transistors operate in an on-off mode, the biasing is not critical, and there is no need for temperature compensation. Some care must be taken to ensure that both transistors are never on at the same time and that provisions exist for stopping the motor when the computer is not operational.

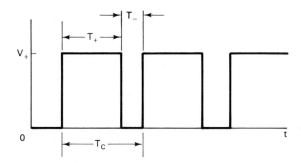

**Figure 4.9**    Pulse-width modulation.

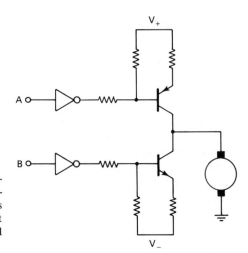

**Figure 4.10** A two-transistor circuit designed for driving a DC motor using pulse-
width modulation (note that the circuits
indicated here as inverters may in fact
be other circuits such as operational
amplifiers).

An additional advantage of using PWM is the simplicity of computer interface, since only two output bits are required.

There are two disadvantages to the use of PWM. First, current control is not possible without significant additional hardware. If direct
control of torque is required, as would be the case, for example, if force
control were needed, then linear operation with an amplifier is pref-
erable to PWM.

A second disadvantage associated with PWM is the electrical noise
which is created by the rapid switching of current through an inductor
(the armature). Proper use of the shielding techniques of Chapter 3 can
eliminate this as a problem.

In summary, PWM provides the quickest and easiest mechanism
for attaining proportional control of a DC motor. However, it lacks
the flexibility required for more sophisticated control strategies such as
control of applied force.

## 4.2 STEP MOTORS

With the continuing *computerization* of electronics, it has become
desirable to have a simple digital positioning device. Such a device
could be controlled by digital pulses on two inputs: move clockwise
and move counterclockwise. The step motor provides one solution to
this problem. In this section we will discuss the step motor and its potential applications. We will use a model for the step motor which is
slightly simpler than realistic motor systems but which presents the basic
concepts.

### 4.2.1 Organization of the Step Motor

Figure 4.11 shows the electrical organization of a three-phase step motor. The armature is wound about an iron rod and is energized with a DC current. A system of slip rings maintains the DC nature of the armature current as the armature rotates. The armature can thus be considered as a simple electromagnet.

At any instant in the operation of a motor, exactly one pair of pole piece windings is energized. Furthermore, each pair (e.g., windings 1a and 1b of Figure 4.11) is wound in such a way that their magnetic fields are collinear. The armature will then rotate until the armature's magnetic field is aligned with the fields of the single pair of pole pieces. This condition is shown in Figure 4.11 with pole pieces 1a and 1b energized. This is a stable condition, and given sufficient time, the motor will come to rest in this position.

If the motor is at rest as shown in Figure 4.11, coil pair 1 is de-energized and, simultaneously, coil pair 2 is energized, then the armature will rotate to bring the armature electromagnet into alignment with coil pair 2. A single "step" of this three-phase motor thus results in a motion of 360/6 = 60 degrees.

By using more pole pieces and by using more complex armature windings, motors may be built with many more steps per revolution; 24 is common.

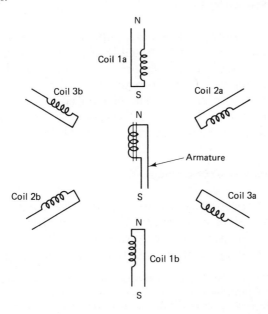

**Figure 4.11**   Schematic organization of a step motor.

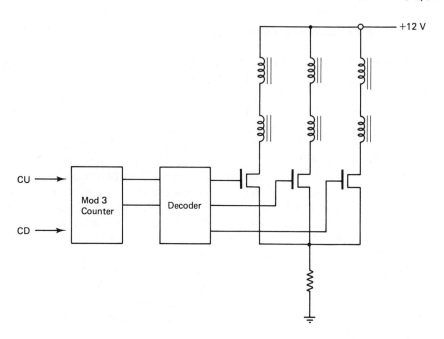

Figure 4.12    Representation of a step motor driver.

### 4.2.2 The Step Motor Driver

Figure 4.12 shows a schematic representation for a step motor driver. A modulo-3 counter followed by a decoder selects one of three possible FET's (field-effect transistors). The selected FET, when energized, allows current to flow through one coil pair.* CU (count up) pulses increment the counter and energize the coils in sequence, resulting in clockwise rotation of the motor. CD (count down) pulses decrement the counter and result in counterclockwise rotation.

### 4.2.3 Performance of Step Motors

To cause a specific clockwise rotation, the computer need only output the appropriate number of CU pulses, resulting in a very simple control scheme. Since no feedback is required, this scheme has found a great number of applications, particularly in printers and plotters.

---

*Figure 4.12 shows field-effect transistors being used to switch the current through fairly high inductance coils. Such switching would result in unacceptably high transient voltages. Hence, commercially available driver circuits are somewhat more complex than shown in this figure. This figure also omits the current reversing logic required to keep the pole alignment correct.

In robotics applications, step motors have found less application, as we shall see.

Figure 4.13 shows the velocity response of a typical step motor running in *single-step mode.* The negative excursions of $\dot{\theta}$ indicate a momentary direction reversal of the motor. This behavior typically results in a high degree of vibration (ringing) in the object being moved. (CU pulse rates of 100 hertz are typical.) The frequency of the ringing is a function of the frictional and inertial loading of the motor. Ringing in single step mode can be reduced by passive *inertial dampers,* which are available from step motor manufacturers.

As the input clock rate is increased, the point will be reached where the next clock pulse occurs before the direction reversal. When running in this mode, the step motor is said to be *slewing.* By slewing, high motor speeds can be accomplished with much less vibration. When a motor is slewing, however, positioning errors can occur. If a motor is moving rapidly, in slewing mode, and the control pulses suddenly stop, the motor may easily have enough forward momentum to carry it all the way past the appropriate stopping point around to the next stable point. In the motor of Figure 4.11, this is an entire revolution. Since the motor has no feedback sensor, it is impossible to detect that this error has occurred.

If the inertia of the load is a constant, then acceleration and deceleration algorithms can be developed that allow step motors to be used at high speed without loss of positional accuracy. Such algorithms are commonly implemented in printers and plotters. Unfortunately, when inertia changes, as it does in robots, the physical performance of the motor system may be faster or slower than that used in the design, and the controller may "lose" steps.

This lack of feedback is both the greatest attribute and the greatest

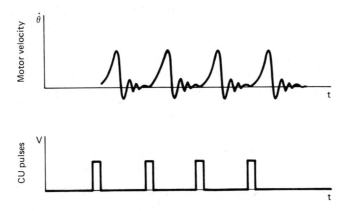

**Figure 4.13**    Response of a step motor (angular velocity versus time).

drawback of the step motor. On the one hand, it provides exceptional simplicity of interface and control, but the range of applications where this control is suitable is quite limited.

An error of one or two steps over a large motion may not be significant, so if the actuator returns to a standard "reset" position fairly often, errors will not accumulate to a measurable extent. The reset position is generally determined by the closing of a switch, introducing a simple degree (1 bit) of feedback.

More sophisticated feedback strategies may be used with step motors, including even force control (Kuo, 1979). However, if feedback needs to be introduced, one is forced to question the use of a step motor at all, since it provides a lower power to weight ratio than a comparable DC motor and may require a controller of comparable complexity.


## 4.3 HYDRAULIC ACTUATORS

Hydraulic systems make use of an incompressible fluid, an oil, which is forced under pressure into a cylinder. The cylinder contains a piston which moves in response to the pressure on the fluid. Both rotary and telescoping (prismatic) actuators are available, and either may be the actuator of choice in robot applications requiring high power. In this section, we will discuss the principles of operation of hydraulic actuators and compare them with electric actuators, in preparation for the discussion of control of such systems in Chapter 5.

### 4.3.1 Principles of Operation of Hydraulic Actuators

The double-acting hydraulic piston is the principal moving part in a hydraulic system. In such a piston (Figure 4.14), fluid can flow into side $A$ and out of side $B$ or vice versa, resulting in movement of the piston to the right or left respectively.

Control over the direction of fluid flow is accomplished by the hydraulic servo valve, detailed in Figure 4.15. A small, high-precision electric motor displaces the valve piston slightly, allowing fluid to flow from the source to the actuator over one hose, returning to the valve over another hose. If displaced in the opposite direction, the valve directs flow in the opposite direction. It should be noted that Figure 4.15 is a simplification of the true construction of a servo valve.

The ideal hydraulic rotary actuator provides shaft torque, $T_m$, proportional to differential pressure, $\Delta P_1$, across the servo valve,

$$T_m = K_P D_m \Delta P_1 \qquad (4.13)$$

# Hydraulic System

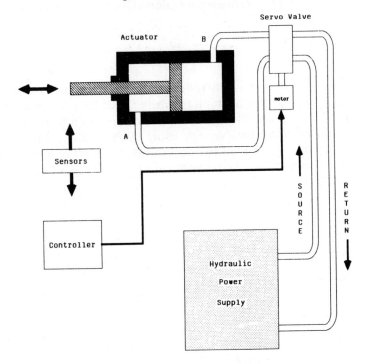

**Figure 4.14**   Hydraulic systems.

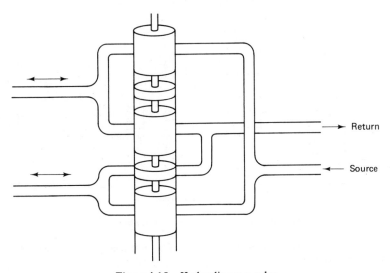

**Figure 4.15**   Hydraulic servo valve.

where $D_m$ is the displaced volume measured in cubic inches. The ideal hydraulic actuator also provides angular velocity, $\dot{\theta}_m$ proportional to flow, $Q$:

$$\dot{\theta}_m = \frac{K_Q}{D_m} Q \qquad (4.14)$$

Either $\dot{\theta}_m$ or $T_m$ may be controlled by use of the appropriate type of servo valve (but not both simultaneously, since they are interrelated). Flow control valves are the most common, and Figure 4.16 shows a block diagram of a control system using such a valve. In Chapter 5, we will discuss the control of hydraulic systems in more detail. At this point, only a few additional comments are required regarding implementational details.

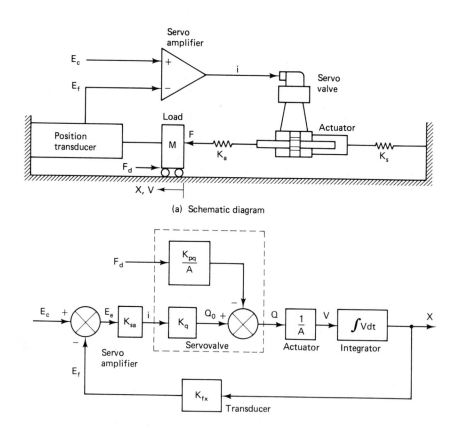

(a) Schematic diagram

(b) Idealized block diagram (dynamic effects neglected)

**Figure 4.16**  (Clark) A hydraulic servo system utilizing position feedback. (*Courtesy MOOG*)

### 4.3.2 Implementation Considerations

Clark (1969) points out that

there are several significant advantages to electrohydraulic servo systems over electric motor drives:

1. Hydraulic drives have substantially higher power to weight ratios, resulting in higher machine frame resonant frequencies for a given power level.
2. Hydraulic drives are much stiffer than electric drives, resulting in higher loop gain capability, greater accuracy, and better frequency response.
3. Hydraulic drives give smoother performance at low speeds and have a wide speed range without special control circuits. They can usually be direct-coupled to the load without the requirement for intermediate gearing.
4. Hydraulic drives are self-cooling and can be operated in a stall condition indefinitely without damage.

While Clark's evaluations are valid, electric motor drives may be more appropriate for certain applications because of the following considerations.

First, it must be emphasized that the hydraulic fluid is incompressible, and when the servo valve is closed, flow cannot occur. This means that when a control system reaches the no-flow condition, it is impossible to backdrive the actuator without incurring physical damage. Thus, compliance with external forces is more difficult to accomplish than with DC motors. It is not impossible; the external forces can be explicitly sensed and the sensory data used to control the actuator. However, this is a fairly complex control technique, as we will learn in Chapter 11.

Second, hydraulic systems are highly nonlinear. These nonlinearities make it more straightforward to implement sophisticated and delicate control using electric drives.

These issues will be discussed in Chapter 5 as they relate to computer control of position and velocity of hydraulic systems and in Chapter 11 as they relate to force control and compliance, important topics for assembly applications.

## 4.4 PNEUMATIC ACTUATORS*

In a pneumatic control system, a compressible fluid, air, is used to drive a piston. As in the case of hydraulic actuators, an electrical signal controls a valve which, in turn, controls the flow to a cylinder. Some of

*The author is grateful to Windell Malpass of Schraeder Bellows Corp. for his assistance in preparation of this section.

the variations in complexity of pneumatic control systems will be discussed in the paragraphs that follow.

In the simplest pneumatic control system, a solenoid opens a valve which allows air flow into a cylinder. The force, $f$, developed by the piston is given by

$$f = pA \qquad (4.15)$$

where $p$ is the supply pressure and $A$ is the cross-sectional area of the piston. To return the piston, the supply valve is closed, and an exhaust valve is opened. With supply pressure eliminated, the piston may be returned by a spring, or, if a double-acting cylinder is used, a constant pressure on the other side will cause a return.

Such simple control is ideal for grippers; the pneumatic piston simply closes the gripper until the gripping force equals the piston force. Unlike a hydraulic gripper, a pneumatic gripper can open and close in response to external forces and hence can *comply* to some extent with unpredictability in its environment.

In addition to their use in grippers, pneumatic components are often used in simple robots. Advocates of pneumatic actuators point to the following advantages:

High speed and relatively high power-to-weight ratio.
Very low cost
Simplicity of control
Noncontamination of work space (unlike an oil leak, a leak in a pneumatic system causes no mess)

A totally pneumatic robot can be sequenced through a complex series of operations by a simple controller which opens and closes valves in order. Such robots normally use fixed stops. That is, a joint moves until its travel is halted by a collision with a rigid piece of metal.

More complex control can be accomplished using variable stops. In variable stop control, a number of mechanical stops exist, which may intercept the joint travel at varying points. The stops are typically engaged by a solenoid and are released by a spring.

Proportional control is possible with pneumatic actuators, although more sophisticated (and, therefore, more expensive) valves are required. Several manufacturers provide $i$ to $p$ transducers that accept a current as input and produce an output pressure directly related to that current. With such transducers, as well as the existence of pressure or flow amplifiers, it becomes feasible to use servo control of a pneumatic joint.

A servo system uses a position sensor to detect true position and compare it with desired position. If an error exists, the servo control moves the joint in a direction to reduce the error. Using a double-acting cylinder or pneumatic motor, with controlled pressure on both sides, servo control is possible. However, at this time only a very few servo-controlled pneumatic robots are in the field.*

Pneumatic servo systems have been seldom used because their dynamic performance is poor in comparison with electrical or hydraulic servos. This poor performance is usually attributed to the compressibility of the fluid and (therefore) to its sluggish time delay. The time delay, $\tau$, for a signal to pass down a pipe of length $l$ is

$$\frac{\tau}{l} = \sqrt{\frac{\rho}{N}} \qquad (4.16)$$

where

$\rho$ = density (kg/cm$^3$)
$N$ = bulk modulus = $\chi P$
$\chi = C_p / C_v = 1.4$
$P$ = absolute pressure

For pneumatic systems at room temperature, this works out to

$$\frac{\tau}{l} = 3 \text{ ms/m}$$

which might be compared with that of a typical hydraulic system:

$$\frac{\tau}{l} \text{ (hyd)} = 0.7 \text{ ms/m}$$

Therefore, propagation speed in a hydraulic system is about four times greater than in a pneumatic system. This difference in speed, while significant, does not explain the inferior performance usually attributed to pneumatic systems when they are compared with hydraulic systems. Mannetje (1981) suggests that the real reason for inferior performance is the use of flow control valves. He recommends that in a pneumatic system, pressure control should be used. He then shows how different strategies in servo design can result in a radical improvement in performance, making pneumatic control viable in many previously unconsidered applications.

## 4.5 SYNOPSIS

### Vocabulary

You should know the definition and application of the terms on the following page.

*Lord Corp.; Pendar, Inc.; International Robomation/Intelligence.

armature
back EMF
compliance
current control
cylinder
DC motor
field
friction
hydraulic system
*i* to *p* transducer
incompressible fluid
inertia
permanent magnet DC motor
pneumatic system
pulse-width modulation
push-pull amplifier
series motor
servo valve
shunt motor
slewing
step motors
torque

**Notation**

| Symbol | Meaning |
|--------|---------|
| $K_t$ | Proportionality constant, relating torque, magnetic flux, and armature current |
| $\Phi$ | Magnetic flux |
| $I_a$ | Armature current |
| $T_m$ | Torque produced by a DC motor |
| $K_f$ | Proportionality constant, relating field current to flux |
| $I_f$ | Field current |
| $J_a$ | Inertia of the armature |
| $F_a$ | Friction of the motor |
| $J_l$ | Inertia of the load |
| $F_l$ | Friction of the load |
| $J_{eq}$ | Total inertial load seen by the motor |
| $F_{eq}$ | Total frictional load seen by the motor |
| $T$ | A torque |
| $\theta$ | Angular position |
| $\dot{\theta}$ | Angular velocity |
| $\ddot{\theta}$ | Angular acceleration |

| Symbol | Meaning |
|--------|---------|
| $E_b$ | Back EMF produced by a motor |
| $V_d$ | Voltage applied to a DC motor |
| $K_P$ | Proportionality constant, relating torque and differential pressure |
| $K_b$ | Proportionality constant, relating back EMF to rotational velocity |
| $K_Q$ | Proportionality constant, relating velocity to flow |
| PWM | Abbreviation for pulse-width modulation |
| $T_c$ | Period of time for one cycle of a PWM driver |
| $T_+$ | Period of time the output of a PWM driver is high |
| $V_{AV}$ | Effective voltage produced by a PWM driver |
| $\Delta P_1$ | Differential pressure across a hydraulic servo valve |
| $D_m$ | Displaced volume |
| $Q$ | Flow |
| $f$ | Force |
| $P$ | Pressure |
| $A$ | Area |
| $\tau$ | Time delay for a signal to travel down a fluid line |
| $l$ | Length of a fluid line |
| $\rho$ | Density of the fluid |

## 4.6 REFERENCES

CLARK, D. *Selection and performance criteria for electrohydraulic servodrives.* 25th Annual Meeting of the National Conference of Fluid Power, October 1969. Also available as *Technical Bulletin 122*, Moog, Inc., East Aurora, N.Y.

KUO, B. *Incremental motion control: Step motors and controls.* Champaign, Ill.: SRL Publishing, 1979.

GOURISHANKAR, V. *Electromechanical energy conversion.* Scranton, Pa.: International Textbooks Company, 1965.

MANNETJE, J. J. Pneumatic servo design method improves system bandwidth twentyfold. *Control Engineering*, pp. 79–83 June 1981.

MILLMAN, J., & HALKIAS, C. *Integrated electronics: Analog and digital circuits and systems.* New York: McGraw-Hill, 1972.

THALER, G., & BROWN, R. *Servomechanism analysis.* New York: McGraw-Hill, 1953.

## 4.7 PROBLEMS

1. A permanent magnet DC motor is tested as follows. With a constant voltage of 15 V applied, and stalled, the motor draws 5 A. When rotating at 50 rad/sec,

the motor draws 1 A.  Use the model of Figure 4.8 and find the effective armature resistance and the back EMF constant, $K_b$.

2. The motor in problem 1, when stalled, draws 5 A and exerts 1 foot-pound of torque. Find $K_{pm}$, as defined in Eq. 4.4.

3. The motor in problem 1 is presented with a voltage step input of 15 V and, unloaded, is measured to accelerate immediately at a rate of 15 m/sec$^2$. Assume a frictionless system and compute the armature inertia.

# 5

---

# *CONTROL*

In this chapter, we will discuss the basic principles of feedback control theory. We will limit the discussion to the control of one joint. Extensions to the coordinated control of several joints will be covered in Chapter 10. Initially, the discussion will focus on the continuous form of control, and differential equations will be formulated that describe the response in time. Then, software control and discrete time will be introduced.

We will assume that we are controlling a rotary joint, since these are very common in robots. The same principles are directly applicable to linear (prismatic) joints, such as pistons.

## 5.1 PROPORTIONAL ERROR CONTROL

The basic principle of control is very straightforward; move the system in the direction that minimizes some error function. An example error function might be $E = \theta_d - \theta$, where $\theta_d$ is the desired angular position and $\theta$ is the actual angular position of the joint. (We will in future sections refer to $\theta_d$ as the *set point*.) When $E = 0$, the joint is at the desired position. If $E$ is negative, then the joint has moved too far and must reverse its motion. Thus, always moving in the direction which makes $E$ approach zero will provide a type of control.

In addition to the drive direction, we should also be concerned with its magnitude. That is, not only must we ask, "In which way should I move the motor?" but also, "How much power (torque) should

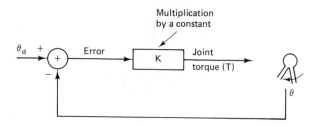

**Figure 5.1**   A PE control system, in which the control torque, $T$, is in proportion to the difference between the desired joint angle, $\theta_d$, and its actual angle ($\theta$).

I apply to the motor?"   Again, the error signal $E = \theta_d - \theta$ provides an answer.   Let us apply a drive signal (a control) which is proportional to $E$. This rule defines a feedback control system as shown in Figure 5.1. Such a system is called a *PE* (*proportional error*) *control* system.

To analyze the performance of PE controllers more carefully, we must have a model of the load being moved.   In Chapter 4, we discussed DC motors and determined that

$$T_m = J_{eq}\ddot{\theta} + F_{eq}\dot{\theta} \qquad (5.1)$$

where $T_m$ is motor torque, $J_{eq} = J_l + J_m$ is the inertia of the load (reflected through the gears) plus the inertia of the motor, $F_{eq}$ is friction, defined similarly, and $\theta$ is angular position.   Henceforth, to simplify notation, we will use $J$ and $F$ to represent $J_{eq}$ and $F_{eq}$, respectively.

We can also determine that, to a reasonable approximation,

$$T_m = K_m I \qquad (5.2)$$

where $I$ is the motor current.   Equating Eqs. 5.1 and 5.2, we find

$$K_m I = J\ddot{\theta} + F\dot{\theta}$$

Now our choice of a control law comes into play.   First, let us assume PE control; that is, we apply a torque proportional to the error signal.   This is accomplished by applying a current proportional to the error.   This assumes, of course, that we have a controllable current source such as the amplifier of Figure 4.7.   We will absorb the proportionality constants together to get

$$K_e(\theta_d - \theta) = J\ddot{\theta} + F\dot{\theta} \qquad (5.3)$$

Without loss of generality, let $\theta_d = 0$.   (The choice of an origin is somewhat arbitrary.)

Now

$$-K_e\theta = J\ddot{\theta} + F\dot{\theta} \qquad (5.4)$$

To see the performance of such a controller, first, suppose that it operates on a frictionless load ($F = 0$).

$$-K_e\theta = J\ddot{\theta} \tag{5.5}$$

This differential equation has as its solution a function that is equal to (minus) its own second derivative. A sine function does nicely, and we predict that a PE controller with no friction will oscillate.

**Example 5.1  Oscillation of a Controller**

Equation 5.5 can be rewritten as $\ddot{\theta} = -\dfrac{K_e}{J}\,\theta$. Assume that $\dfrac{K_e}{J} = 100$. Determine the solution to Eq. 5.5, and, from that solution, determine the frequency of oscillation.

**Solution:**   Equation 5.5 describes a function proportional to its own second derivative. A sine is one such function, as is a cosine. Without information about initial conditions, we cannot determine which, or what the phase will be. Thus, we choose

$$\theta = \sin 10t.$$

We can verify that this satisfies Eq. 5.5 by differentiation.

$$\dot{\theta} = 10 \cos 10t$$

$$\ddot{\theta} = -100 \sin 10t = -100\theta$$

The motor so controlled will oscillate at a 10-Hz rate.

In the presence of friction, Eq. 5.4 describes the behavior and has a solution of

$$\theta = \exp\left(-F/2J\ t\right)[C_1 \exp\left(\tfrac{1}{2}\ \omega t\right) + C_2 \exp\left(-\tfrac{1}{2}\ \omega t\right)]$$

where

$$\omega = \sqrt{\left(\frac{F^2}{J^2}\right) - \left(\frac{4K_e}{J}\right)} \tag{5.6}$$

The *damping term*, $\exp\left(-F/2J\ t\right)$ guarantees that, with increasing time, the joint will get closer and closer to its goal of $\theta = 0$. Furthermore, we can see that if $F^2/4K_e > J$, then the term under the radical will be positive, resulting in even more damping. Such a solution is referred to as *overdamped*, and is demonstrated in Figure 5.2.

If $F^2/4K_e < J$, then the exponent is complex and the solution is a damped sinusoid, shown in Figure 5.3. Intuitively, these solutions make sense, for high friction means "hard to start, easy to stop"; and high inertia means "hard to start, hard to stop."

If $F^2/4K_e = J$, the solution is *critically damped*. That is, it gets to the goal as quickly as possible without overshoot.

Clearly, by choosing $K_e$, we can affect the performance of the system. We will discuss this at a greater length later.

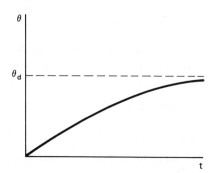

**Figure 5.2**   Overdamped.

There are some problems with the proportional error control system when we must hold a load against gravity. To do so requires a torque, so we cannot hold against gravity without an error, since no error would imply no torque. This is known as the *steady-state error* problem.

A second problem with proportional error control is overshoot. That is, a manipulator operating under control of a proportional servo has only friction to slow it down. To see this, suppose that the arm is close to its destination. Then

$$T = K_e(\theta_d - \theta)$$

is quite small, but not negative. If there is much friction, the arm may stop short of $\theta_d$ due to lack of drive, but if friction is small and inertia is large (relatively), then the arm may move on past $\theta_d$. Now the error signal is negative, torque is backward, and the arm will be driven back to $\theta_d$. In the meanwhile, however, it has *overshot* its goal. If the task to be performed requires critical positioning, as in moving television picture tubes, for example, the occurrence of overshoot can be disastrous.

Let us now consider some techniques for dealing with steady-state error and overshoot.

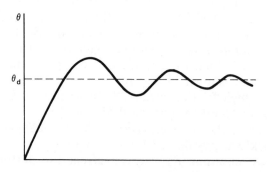

**Figure 5.3**   Underdamped.

## 5.2 THE STEADY-STATE ERROR PROBLEM

To hold a load against gravity, a controller must exert a torque (or force). To do so with a PE controller requires an error, as the following example illustrates.

### Example 5.2  Effect of Loading on a PE Controller

A robot with one prismatic joint is oriented so that the joint is aligned vertically. The actuator is controlled by a controller that relates force (measured in kilograms) to position (measured in centimeters) by

$$f = K_e(y_d - y)$$

where

$$K_e = 0.6 \text{ kg/cm}$$

Assume that the load seen by the actuator (including the mass of the joint itself) is 30 kg. If the desired vertical position is $y = 300$ cm, determine the true position in the steady state (that is, no acceleration and no velocity).

**Solution:**

$$y_d - y = \frac{f}{K_e} = \frac{30 \text{ kg}}{0.6 \text{ kg/cm}} = 50 \text{ cm}$$

Since $y_d = 300$ cm, $y$ must be 250 cm.

Thus, to hold a load of 30 kg against gravity, this controller will be in error by 50 cm, in the steady state.

One approach to dealing with steady-state error is to produce as output a torque $T = L + K_e(\theta_d - \theta)$, where $L$ is a constant sufficient to hold the load when $\theta_d - \theta$ is zero.

Use of this approach requires that the load be known precisely. In the case of robots, this knowledge is difficult to achieve since the load on a particular joint is usually a function of the positions and motions of the other joints.

An alternative is to make the drive signal equal to the integral of the error with respect to time. That is, allow the output of the servo (the motor torque) to accumulate with time and make the rate of accumulation proportional to the error signal. Such a controller essentially finds the constant $L$ defined earlier by the experimental technique of increasing $L$ slowly until the load can be held stationary with no error.

A robot operation with such a *PI (proportional integrating)* controller on its vertical axis can be observed to droop when a load is suddenly applied and then rise back to the desired position as the integral term builds up.

Of course, the improved performance of an integrating controller does not come for free. We have been discussing how the integrator helps the steady-state error problem. That is, integration is of assistance when the arm is stationary or moving slowly. When an integrating controller is used to achieve fast motion, however, it tends to increase the overshoot. In fact, under certain loading conditions, such controllers can be unstable and oscillate about the desired point. Thus, while reducing one problem, steady-state error, we have made another problem, overshoot, even worse. Such are the joys of engineering (or economics, for that matter)!


### 5.3 THE OVERSHOOT PROBLEM

Overshoot occurs because the controller has an insufficient mechanism for "applying the brakes." A PE or PI controller in fact has nothing to stop the arm other than friction. If there is any positive error at all, $\theta_d - \theta$ is positive, and the motor will have positive (although small) drive applied right up to the point where the error goes to zero. If friction is small and inertia large relative to the friction, a joint driven by such a controller will overshoot.

To provide a degree of active braking, we can use the following concept:

1. If error is large (we are a long way from the goal) and the velocity is small, apply a large drive.
2. If error is small (we are close to the goal) and the velocity is high, apply a negative drive.

The simplest way to achieve this is to make the drive torque proportional to the derivative (rate of change ) of position with respect to time:

$$T = K_e(\theta_d - \theta) - K_d\dot{\theta} \qquad (5.7a)$$

where $\dot{\theta}$ is the angular velocity.

This equation defines the operation of a *PD (proportional derivative) controller.*

The ability of this controller to handle overshoot then depends on the gains of the controller, $K_e$ and $K_d$, and the inertia and friction of the load. One cannot always guarantee zero overshoot unless something is known about maximal values for inertia. Choice of optimal $K_e$ and $K_d$ is then possible. However, these constants are most often determined experimentally. Increasing $K_d$ is equivalent (for purposes of control) to increasing the friction of the system.

To see this, we once again equate Eqs. 5.1 and 5.2 and substitute the new control law

$$K_e(\theta_d - \theta) - K_d\dot{\theta} = J\ddot{\theta} + F\dot{\theta} \qquad (5.7b)$$

again assuming without loss of generality that $\theta_d = 0$

$$-K_e\theta - K_d\dot{\theta} = J\ddot{\theta} + F\dot{\theta} \qquad (5.8)$$

Rearranging terms yields

$$-K_e\theta = J\ddot{\theta} + (F + K_d)\dot{\theta} \qquad (5.9)$$

which is exactly Eq. 5.4 with $K_d$ added to the friction. Thus, a PD controller has exactly the same behavior as a PE controller, but the designer now has another parameter to adjust for best performance, a parameter that functions exactly like friction.

A controller with derivative feedback can be combined with the concept of integration to yield a *PID* (*proportional integral derivative*) *controller.* There are several ways in which one could configure such a controller. One such configuration is shown in Figure 5.4.

The torque provided by a PID controller satisfies

$$T_m(t) = K_e(\theta_d - \theta(t)) + K_i \int_{t_0}^{t} (\theta_d - \theta(\tau))\,d\tau - K_d\dot{\theta}(t) \qquad (5.10)$$

As we will discuss in the next section, this equation could be written in discrete form as

$$T_m(i) = K_e(\theta_d - \theta(i)) + \frac{K_i}{i}\sum_{l=0}^{i}(\theta_d - \theta(l)) - K_d\dot{\theta}(i)$$

at time $t = i$, which explicitly indicates the use of discrete time computations.

A PID controller trades off the possibility of overshoot against the

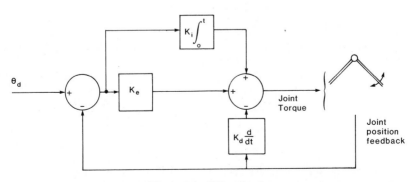

**Figure 5.4**   PID control.

speed of the joint motion. Increasing $K_d$ tends to slow the arm down since it increases the negative contribution to torque due to velocity. Decreasing $K_d$ decreases the damping of the system, thus increasing the likelihood of overshoot.

Because of their relatively simple implementation and *robustness* (ability to adapt to changing loads), PID controllers are probably the most commonly used controllers today, even though their performance is not necessarily optimal.

## 5.4 THE SAMPLED-DATA CONTROLLER

The control system of Figure 5.4 could easily be implemented with operational amplifiers, as shown in Figure 5.5. Proper placement of capacitors provides differentiation and integration, and use of potentiometers provides control of the gains $K_e$, $K_d$, and $K_i$. Such analog circuits were the standard means for realizing servo controllers until very recently. It should be noted that the differentiator using op-amps shown in Figure 5.5 tends to be extremely sensitive to noise. A superior practice is to sense $\dot{\theta}$ directly, with a tachometer, for example, rather than to differentiate $\theta$. However, silicon technology now provides means for performing the same operations digitally and thus avoids the difficulties of accuracy, drift, and temperature compensation that plague analog circuits. Furthermore, the continuing decrease in the cost of digital circuitry makes the digital approach increasingly attractive.

Probably the most obvious way in which to implement a PID controller digitally is to replace each block of Figure 5.4 with its digital

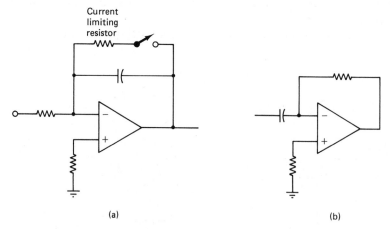

(a)                                                                    (b)

**Figure 5.5(a)** Integrator using operational amplifier. The switch is used to set initial conditions to zero. **(b)** Differentiator using operation amplifier.

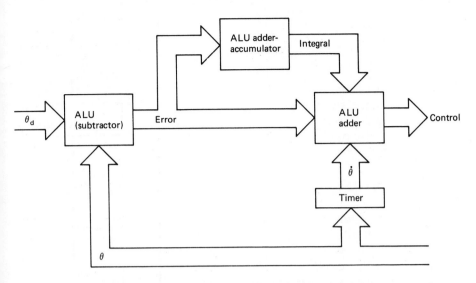

**Figure 5.6**    Digital implementation of PID control.

equivalent, as shown in Figure 5.6.  There, an ALU (arithmetic-logic unit) is used to compute the error signal; an ALU plus accumulator provides the integration; and a counter, with timer, provides differentiation. We will refer to this as a *parallel digital implementation.*

Such an approach is far from cost effective when compared with performing the same operations in a computer.  The cost of microprocessors is now so low as to make a software approach far more economical, as well as providing the flexibility of easy modification. Although the parallel approach might be necessary if the variable to be controlled changes very rapidly (e.g., a new and significantly different value every 10 microseconds), the joint variables of a robot change much more slowly than this.  Later in this section, we will examine the speed requirements for a computer controller.

Figure 5.7 shows a flow chart representing software implementation of a PE controller. The unique difference between a software controller such as this and a continuous time controller is the effect of *loop cycle time*.  Loop cycle time refers to the amount of time required from when the input is read until the input is read again. The functioning of such a controller is analogous to riding a bicycle with one's eyes closed. The rider takes a quick look, determines that he still has far to ride, closes his eyes, and pedals hard.  Sometime later, he looks again, determines that he is now closer, closes his eyes, and does not pedal quite as hard.  Obviously, the performance of the system depends strongly on how often the rider takes a look, or on how often control returns to block 1 of the flow chart.

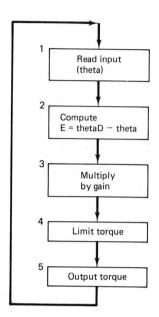

**Figure 5.7** Flow chart of a PE controller.

The minimum loop cycle time (or *sampling interval*) will vary from one robot to another. It depends on the mechanical time constant of the physical system, which is related to the inertia. Larger systems have a longer mechanical time constant. Studies (Paul, 1982) have shown that the minimal value lies between 5 and 100 ms. Many robot systems use 16 ms (1/60 sec) since the line provides such convenient source of timing signals.

## 5.5 THE VOLTAGE-CONTROLLED DC MOTOR

If a permanent magnet DC motor (pmDC) is driven by a controllable current source, such as the circuit shown in Figure 4.7, then we can use Eq. 5.3 to describe the system's performance. It is not always possible, however, to supply a controlled current. If we instead control the voltage to the motor, we must take a closer look to analyze the system.

Figure 4.8 represents a pmDC motor in the steady state, that is, turning at a constant velocity. In this model, we ignore the motor's inductance. This simplification is made since transient effects due to inductance are, in general, much faster than the mechanical actions we are controlling. Thus, we assume that the physical motor cannot respond to the inductive transients.

If we happen to be controlling voltage, with a PD controller, we have

$$V_d = K_e(\theta_d - \theta) - K_d \dot{\theta} \tag{5.11}$$

Then, the current through the motor is

$$I = \frac{V_d - E_b}{R} \tag{5.12}$$

$$= \frac{1}{R} [K_e(\theta_d - \theta) - K_d \dot{\theta} - E_b] \tag{5.13}$$

Torque is related to current by

$$T_m = K_m I_a \tag{5.14}$$

$$= \frac{K_m}{R} [K_e(\theta_d - \theta) - K_d \dot{\theta} - E_b] \tag{5.15}$$

The back EMF of the motor results from its acting as a generator and is proportional to rotational velocity:

$$E_b = K_b \dot{\theta} \tag{5.16}$$

So Eq. 5.15 becomes

$$T_m = \frac{K_m}{R} [K_e(\theta_d - \theta) - K_d \dot{\theta} - K_b \dot{\theta}]$$

$$= \frac{K_m}{R} [K_e(\theta_d - \theta) - (K_d + K_b)\dot{\theta}] \tag{5.17}$$

$$T_m = K_1(\theta_d - \theta) - K_2 \dot{\theta}$$

We see that this equation has exactly the same form as Eq. 5.7a and thus will have the same type of solutions. However, the damping constant has been increased by $K_b$. Hence, the effect of the back EMF of the motor on a voltage-driven controller is to increase the damping constant of the controller.

## 5.6 CHOOSING SERVO GAINS

As we saw in Eq. 5.10, the torque from a PID controller is described by

$$T = K_e(\theta_d - \theta) + K_i \int (\theta_d - \theta) \, dt - K_d \dot{\theta} \tag{5.10}$$

As before, we could equate this torque to the mechanical torque, $T = J\ddot{\theta} + F\dot{\theta}$, and develop a differential equation to characterize the performance. In this case, the differential equation would be third order. Nonetheless, conditions could be developed for best performance and formulated as equations involving the gains $K_e$, $K_i$, and $K_d$ and the loads $J$ and $F$. In Chapter 10, we will do just that and discover that since $J$ and $F$, $J$ especially, change radically with arm configuration, such simple optimization techniques are doomed to failure.

### The $\theta$-$r$ Manipulator

To see the difficulties inherent in choice of gains, we will consider the *$\theta$-$r$ manipulator*. This simple robot, shown in Figure 5.8, has only two actuators, one driving a rotary joint and one driving a prismatic joint. If we are using PD control on the rotary joint, then the torque from the controller is

$$T = -K_e\theta - K_d\dot{\theta}$$

and the torque due to the motion of the load is

$$T = J\ddot{\theta} + F\dot{\theta}$$

Equating these, as earlier, we once again find Eq. 5.9, with time solution

$$\theta = \exp - \frac{F + K_d}{2J} t \, [C_1 \exp (\omega t/2) + C_2 \exp (-\omega t/2)] \quad (5.18)$$

where

$$\omega = \sqrt{[(F + K_d)^2/J^2] - (4K_e/J)} \quad (5.19)$$

The critically damped solution gives the most speed without overshoot and can be formed by setting the term under the radical equal to zero.

$$(F + K_d)^2 - 4K_eJ = 0 \quad (5.20)$$

Any choice of $K_e$ and $K_d$ satisfying this condition will be critically damped.

Two observations are in order. First, the *critically damped solution* is in fact many solutions. Any of an infinite number of choices of $K_e$ and $K_d$ can satisfy Eq. 5.20 as is true any time we have two variables and one equation. To find a unique choice for $K_e$ and $K_d$ requires adding another constraint. One such constraint which is popular is to require that some function be minimized, such as

$$\min \int u^2 \, dt$$

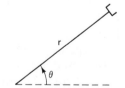

**Figure 5.8**   $\theta$-$r$ manipulator.

where $u$ is the output of the controller.

$$u = K_e\theta - K_d\dot{\theta}$$

This is referred to as a *minimum energy controller*. Use of this condition can result in a unique choice of $K_e$ and $K_d$. Details of how this is done are beyond the scope of this book. The reader is referred to basic texts in optimal control (Bryson and Ho, 1969).

The second observation that needs to be made in this context is the dependence of the solution on $J$ and $F$, as $r$ varies.

The inertial load seen by a rotating actuator is

$$J = mr^2 \tag{5.21}$$

where $m$ is the effective mass and $r$ the effective radius. That is, $m$ is the mass at the end of a massless rod of length $r$.

Substituting $J = mr^2$ into Eq. 5.9, we find

$$mr^2\ddot{\theta} + (F + K_d)\dot{\theta} + K_e\theta = 0 \tag{5.22}$$

This equation still has the solution given by Eq. 5.18 and can be critically damped if

$$(F + K_d)^2 - 4mK_er^2 = 0 \tag{5.23}$$

If $r$ is known and constant, we have no problem. However, now let us consider two more difficult (and more common) cases: constant gains but different $r$ and coordinated motion.

### Constant Gains

If we choose a $K_e$ and $K_d$ to satisfy Eq. 5.20, we have a solution that is good (critically damped) at only one point. We see from Eq. 5.23 that if $r$ takes on a different value, our previous solution will be incorrect. In fact, if $r$ becomes smaller, the solution will be overdamped and if $r$ becomes larger, it will be underdamped. (The proof is left as an exercise.) So with constant gains, the system is virtually always performing poorly. Of course, for any constant value of $r$, Eq. 5.23 can be used to find good values for the gains, and this technique is often used.

### Coordinated Motion

If both joints move simultaneously, the situation is much more complex, for in that case, $r$ and, therefore, $J$ are functions of time. The system must be described by more sophisticated modeling techniques which incorporate the interaction of forces. This topic will be discussed in more detail in Chapter 10.

For now, let us conclude by suggesting that the gains be chosen experimentally by making adjustments and observing the performance. Two rules of thumb are

1. Increasing $K_d$ leads to more sluggish, overdamped response.
2. Increasing $K_i$ significantly increases the likelihood of overshoot and oscillations.

## 5.7 CHOOSING THE CONTROLLED VARIABLE

Up to this point, we have discussed only control of angular position, $\theta$. However, there is no reason that other parameters could not be controlled. In general, we must be concerned with two variables, the variable that is implicitly controlled and that which is explicitly controlled. For example, in a DC motor, we implicitly control torque, and in a hydraulic system, we implicitly control velocity. The variable explicitly controlled is the variable that creates the error signal. For example, we may feed back true velocity, subtract it from desired velocity, and use that error signal to provide the drive signal, thus implementing a velocity servo. In this context, we will use the term *controlled variable* to mean that variable that is fed back and subtracted.

As system designers, we have the freedom to choose the controlled variable according to the application at hand. For example, we may need to avoid abrupt starts and stops in a very-high-gain hydraulic system and thus may program in an acceleration algorithm, as shown in Figure 5.9. One convenient means for implementing acceleration is to provide input to a velocity servo.

Although we have the freedom to choose the controlled variable,

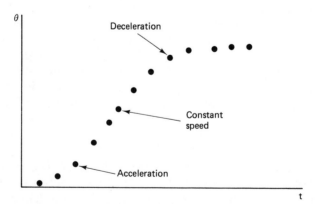

**Figure 5.9**  Set points for a position servo.  By changing the set point as a function of time, acceleration may be controlled.

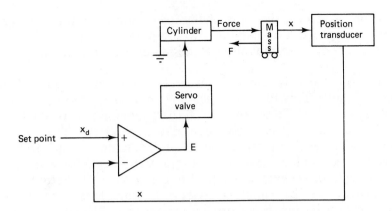

**Figure 5.10**    A hydraulic position control system.

the physical nature of the system may influence our choice of which variable to choose. For example, in a flow-controlled hydraulic system, the servo valve controls flow and, therefore, velocity. We can make use of this fact to implement simple acceleration algorithms. Furthermore, if velocity is not the variable we intend to control, the fact that it is the implicitly controlled variable may lead to unexpected results. To see what these may be, let us consider position control of a hydraulic system.

Figure 5.10 shows the structure of a position controller. Figure 5.11 is derived from Figure 5.10, neglecting dynamic characteristics of both the servo valve and the actuator. The model does include, however, the frictional (damping) effects within the actuator and load friction. From Figure 5.11, we see that the servo valve produces an output flow or, if we neglect damping ($F = 0$), a flow proportional to the error in position. Hence, we have a PE controller driving a system with damping. But unlike the PE controller we studied earlier, this error signal results in a velocity, not a torque.

In Figure 5.11, the output of the actuator is seen to be a velocity. The load is then modeled as a simple integrator that converts velocity into position. Such a simplistic model for the physical world is unreal-

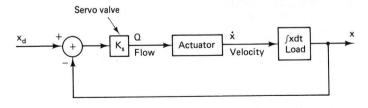

**Figure 5.11**    Representation of Figure 5.10 ignoring dynamics.

istic in that it ignores the mass of the load. However, for sufficiently powerful actuators, this may be an acceptable model. (See Clark, 1969, for a serious introduction to design of electrohydraulic control systems.)

The fact that it is velocity that is ultimately controlled leads to a surprising result: This PE controller does not suffer from the steady-state error problem. To see this, recall that any nonzero error will leave the servo valve at least partially open, resulting in flow. That flow will continue, with resultant displacement of the actuator, until true zero error is reached. In fact, the velocity-to-position integration acts exactly like the integrator we explicitly added to create the PID controller.

Thus, we see that simple analysis of a hydraulic system indicates that PE control results in performance similar to a PID controller in an electrical system. The integration results from the velocity control inherent in the servo valve and the damping from the high friction in the oil seals. Similar nonquantitative analysis of other control systems can likewise provide insight into perforamce. The reader is referred to Clark (1969) for both insight and first-level quantitative analysis on not only position, but also velocity and pressure controllers.

## 5.8 SYNOPSIS

### Vocabulary

You should know the definition and application of the following terms:

critically damped
friction
inertia
loop cycle time
overdamped
overshoot
proportional derivative control
proportional error control
proportional integral derivative control
robustness
sampling interval
set point
steady-state error
torque
underdamped

### Notation

| Symbol | Meaning |
| --- | --- |
| $\theta$ | Angular position, joint variable for a typical rotary joint |
| $\dot{\theta}$ | Angular velocity |
| $E$ | Position error |
| $\theta_d$ | Desired position |
| $J_{eq}$ | Inertia seen by the actuator (including its own inertia) |
| $F_{eq}$ | Friction seen by the actuator (including friction due to the actuator itself) |
| $J$ | Same as $J_{eq}$ |
| $F$ | Same as $F_{eq}$ |
| $T_m$ | Torque of a DC motor |
| $T$ | An arbitrary torque |
| $I$ | Current into a DC motor |
| $K_m$ | Proportionality constant, relating current and torque |
| $K_e$ | A servo gain (a constant multiplying the error term) |
| $K_d$ | A servo gain (a constant multiplying the derivative term) |
| $K_i$ | A servo gain (a constant multiplying the integral term) |
| $V_d$ | Voltage applied to a DC motor |
| $R$ | Resistance of the armature of a DC motor |
| $E_b$ | Back EMF generated by a DC motor |
| $K_b$ | Proportionality constant, relating back EMF to angular velocity |
| $K_1$ | A proportionality constant, equal to $K_m K_e / R$ |
| $K_2$ | A proportionality constant, equal to $K_m(K_d + K_b)/R$ |
| $r$ | The second joint of the $\theta$-$r$ manipulator |
| $\theta$ (alternative) | The first joint of the $\theta$-$r$ manipulator |
| $u$ | The output of an arbitrary controller |
| $m$ | A mass |
| $Q$ | Flow, quantity controlled by a hydraulic servo valve |

## 5.9 REFERENCES

BRYSON, A., and HO, Y. *Applied optimal control.* Waltham, Mass.: Blaisdell, 1969.

CLARK, D. *Selection and performance criteria for electrohydraulic servodrives.* 25th Annual Meeting of the National Conference on Fluid Power, October 1969. Also available as *Technical Bulletin 122,* Moog, Inc., East Aurora, N.Y.

PAUL, R. P. *Robot manipulators.* Cambridge, Mass.: M.I.T. Press, 1982.

## 5.10 PROBLEMS

1. In a permanent magnet DC motor, write the equation that relates *voltage* applied to the motor to the torque produced.
2. Write the equation that relates the torque applied to a physical system to angular position. Assume that the only loads on the system are inertia and gravity and that the system rotates in a horizontal plane.
3. Write the equation that relates the voltage out of a PD controller to the inputs (set point, true position, and velocity).
4. Rework problems 2 and 3, assuming that the motor rotates in a vertical plane.
5. Use the results of problems 1, 2, and 3 to write a differential equation that will predict the behavior of the system so described.
6. Use the results of problems 1, 2, and 4 to write a differential equation of the system so described.
7. What is the fallacy in trying to use a simple differential equation such as that obtained in problem 5 to predict the performance of the various joints of a robot?

# 6

---

# ROBOT COORDINATE SYSTEMS

In Chapters 6, 7, and 8, we will discuss the role of the computer in resolving relationships, particularly relationships between positions and velocities. These may be positions and velocities in *Cartesian space*, the normal measurement space in which the robot sits, or *joint space*, an ordered six-tuple describing the angular position of each joint.

In this chapter, we will develop a consistent and useful notation for representing positions and velocities in Cartesian space. In Chapter 7, on kinematics, we will learn how to relate such descriptions to joint space.

## 6.1 ON POSITION AND ORIENTATION

To perform the full range of manipulation tasks, a robot must not only be able to reach any point in its working space, it must be able to reach that point at any arbitrary orientation. For example, to thread a bolt, the robot must not only grasp the bolt, it must also hold the bolt in a controlled orientation (Figure 6.1). The three variables $\langle x, y, z \rangle$ (or $\langle \theta, \phi, r \rangle$ or other) that define the position of the hand are not adequate to also describe the orientation of the hand. Three more variables will be required. These are usually rotary variables and are most often referred to as *roll*, *pitch*, and *yaw*.

To understand the meaning of these terms, hold your arm out, with wrist unbent and fingers straight (Figure 6.2). Now rotate your hand from palm down to palm up, while keeping the fingers pointed in

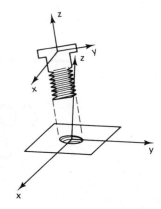

**Figure 6.1** To thread a bolt successfully, the robot holding the bolt must align the z axes.

the same direction. That motion is roll. Next, keep the forearm straight, and without rolling the wrist, move the direction of the hand from straight out to down and back to up (wave!). That motion is pitch. Finally, with the fingers once again pointed straight out, and with no roll or pitch, point the fingers as far right as possible (about 45 degrees for the right hand for the average human wrist) and then as far left as possible. That final motion is yaw.

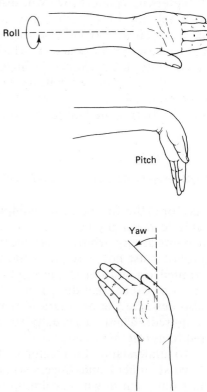

**Figure 6.2** Roll, pitch, and yaw.

These independent motions are referred to as *degrees of freedom*, and six are required to reach an arbitrary point at an arbitrary orientation. The human arm has exactly six, two in the shoulder, one in the elbow, and three in the wrist. Consequently, even without moving the torso, people can point their hands in any direction, at any point.

One comment needs to be made with regard to the statement that six degrees of freedom provide complete flexibility. It is true only if complete freedom of motion is available for each joint. In general, this is not the case. For example, the yaw capability of the human wrist is only about 60 degrees. Hence there are some orientation/position combinations that are impossible for humans. The same will be true for robots.

Six degrees of freedom requires six actuators, and these are generally independent motors. However, because a robot has six motors does not mean that it has six degrees of freedom. For example, finger open/close is not considered a degree of freedom since it does not help position the robot.

In this chapter, we will examine a method for representing position and orientation of a point in space and relate that representation to robot motions.

## 6.2 COORDINATE SYSTEMS—RELATIVE FRAMES

Consider the problem of a robot holding a part for insertion into several numerically controlled machines for various operations such as drilling and grinding. The robot first grasps the part in a specified way and inserts it into the first machine. After the machining operation, the robot grasps the part in a different way and inserts it into the second machine. The problem is, how to specify exactly those two gripping positions. We might say, "Grasp the part at a point 10 inches from the end for the first operation and 15 inches from the end for the second." That is, we are specifying the grasp position relative to a coordinate system defined on the part.

In the following discussions, we will frequently use the word *configuration*. In using this word, we mean both the three-dimensional position and the orientation. Hence, six numbers are required to specify a "configuration".

The grasp points are specified relative to a coordinate system defined on the part. But the location of the part itself (and, hence, its coordinate system) is a variable, known at any instant in time relative to some other coordinate system. Thus, we must learn how to solve the following problem:

Given a location, *A*, which is known in a coordinate system, *B*,

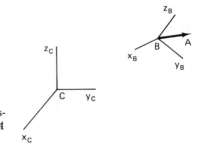

**Figure 6.3**  Coordinate system $B$ is displaced and rotated from system $C$. $A$ is known in $B$. Where is $A$ in $C$?

and given that the location of the coordinate system $B$ is known in another coordinate system, $C$, find the location of the point $A$ as measured in coordinate system $C$ (Figure 6.3). To see this most clearly, let us temporarily ignore the complication due to rotation and consider only displacements. In this case, the coordinate axes of $B$ are parallel to $C$, and $A$ becomes a single point. Thus, $A$ may be described by a vector

$$A = [^B x_A, {}^B y_A, {}^B z_A]^\mathsf{T} \tag{6.1}$$

The superscript on the left is used to denote that this variable is measured with respect to the coordinate system named $B$. Furthermore, the origin of coordinate system $B$ is known relative to $C$:

$$B = [^C x_B, {}^C y_B, {}^C z_B]^\mathsf{T} \tag{6.2}$$

(The transpose notation is used since these vectors should be written as column vectors, but for typographical convenience, we use row vector notation.)

Then $A$, relative to $C$, is

$$^C A = {}^B A + {}^C B \tag{6.3}$$

$$= [^C x_B + {}^B x_A, {}^C y_B + {}^B y_A, {}^C z_B + {}^B z_A]^\mathsf{T} \tag{6.4}$$

Thus, vector addition provides the solution to such problems, provided there is no rotation. As we add to this notation the possibility of rotation, we need to define the concept of a frame.

A *frame* is a representation for a coordinate system so that the representation includes the possibility that the coordinate system may be displaced (translated) and/or rotated with respect to another coordinate system. It must be emphasized that the concept of a frame is not meaningful in itself, but only as it may be related to another coordinate system (frame). It is quite literally a *frame of reference*. That other coordinate system may be explicitly defined as was $B$ in the preceding example, or it may be implicit, perhaps some *universe frame* that acts as a global origin for all elements under consideration.

Before we can relate points relative to different coordinate systems in full generality, we need to learn what happens to a point when it is both rotated and translated relative to a fixed coordinate system.

## 6.3 ROTATIONS

Given a point $[^A x, \,^A y, \,^A z]^T$ known relative to a coordinate system named $A$, we can find a new point $[x', y', z']^T$ that results from rotating the point about the $z$ axis of $A$ through an angle $\theta$, by multiplying by a *rotation matrix*

$$
\begin{bmatrix} x' \\ y' \\ z' \end{bmatrix} = \begin{bmatrix} \cos\theta & -\sin\theta & 0 \\ \sin\theta & \cos\theta & 0 \\ 0 & 0 & 1 \end{bmatrix} \begin{bmatrix} ^A x \\ ^A y \\ ^A z \end{bmatrix} \qquad (6.5)
$$

### Example 6.1  Rotation

Find the result of rotating the point $[7, 2, 5]^T$ through $90°$ (Figure 6.4) in the positive direction about the $z$ axis.

**Solution:**    Substitute into Eq. 6.5

$$
\begin{bmatrix} -2 \\ 7 \\ 5 \end{bmatrix} = \begin{bmatrix} 0 & -1 & 0 \\ 1 & 0 & 0 \\ 0 & 0 & 1 \end{bmatrix} \cdot \begin{bmatrix} 7 \\ 2 \\ 5 \end{bmatrix}
$$

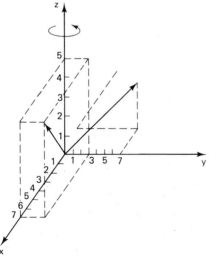

**Figure 6.4**  Effect of rotating a point 7, 2, 5 through $90°$ about the $z$ axis.

We can likewise define 3 × 3 matrices representing rotation about the $x$ and $y$ axes. However, before doing so, we introduce the following coordinate notation.

## 6.4 HOMOGENEOUS COORDINATES

We will want to describe a generalized transformation by a single matrix that combines the effects of translation and rotation. To accomplish this, we first augment the definition of a vector by adding a 1.

$$\mathbf{V} = \begin{bmatrix} x \\ y \\ z \\ 1 \end{bmatrix} \tag{6.6}$$

We then redefine the rotation matrices, making them 4 × 4, by adding a row and a column:

$$\text{Rot}(x, \theta) = \begin{bmatrix} 1 & 0 & 0 & 0 \\ 0 & \cos\theta & -\sin\theta & 0 \\ 0 & \sin\theta & \cos\theta & 0 \\ 0 & 0 & 0 & 1 \end{bmatrix} \tag{6.7}$$

$$\text{Rot}(y, \theta) = \begin{bmatrix} \cos\theta & 0 & \sin\theta & 0 \\ 0 & 1 & 0 & 0 \\ -\sin\theta & 0 & \cos\theta & 0 \\ 0 & 0 & 0 & 1 \end{bmatrix} \tag{6.8}$$

$$\text{Rot}(z, \theta) = \begin{bmatrix} \cos\theta & -\sin\theta & 0 & 0 \\ \sin\theta & \cos\theta & 0 & 0 \\ 0 & 0 & 1 & 0 \\ 0 & 0 & 0 & 1 \end{bmatrix} \tag{6.9}$$

Finally, we can combine the effects of translation and rotation by placing the displacement vector in the fourth column:

$$\text{Trans}(a, b, c) = \begin{bmatrix} 1 & 0 & 0 & a \\ 0 & 1 & 0 & b \\ 0 & 0 & 1 & c \\ 0 & 0 & 0 & 1 \end{bmatrix} \qquad (6.10)$$

We see that matrix multiplication using Trans $(a, b, c)$ is equivalent to addition of a 3-vector:

$$\begin{bmatrix} a \\ b \\ c \end{bmatrix} + \begin{bmatrix} d \\ e \\ f \end{bmatrix} = \begin{bmatrix} a+d \\ b+e \\ c+f \end{bmatrix} \qquad (6.11)$$

and

$$\begin{bmatrix} a+d \\ b+e \\ c+f \\ 1 \end{bmatrix} = \begin{bmatrix} 1 & 0 & 0 & a \\ 0 & 1 & 0 & b \\ 0 & 0 & 1 & c \\ 0 & 0 & 0 & 1 \end{bmatrix} \cdot \begin{bmatrix} d \\ e \\ f \\ 1 \end{bmatrix} \qquad (6.12)$$

The combination of rotations and translations can likewise be accomplished by a matrix multiplication.

**Example 6.2  Combined Motions**

Find the result of acting on a point, $[a, b, c]^{\mathsf{T}}$ by

    1. A rotation of $90°$ about the $z$ axis, followed by
    2. A rotation of $90°$ about the $x$ axis, followed by
    3. A translation of $[d, e, f]^{\mathsf{T}}$

**Solution:**

$$[\text{Trans}(d, e, f)] \cdot [\text{Rot}(x, 90)] \cdot [\text{Rot}(z, 90)] \cdot \begin{bmatrix} a \\ b \\ c \\ 1 \end{bmatrix}$$

$$= \begin{bmatrix} 1 & 0 & 0 & d \\ 0 & 1 & 0 & e \\ 0 & 0 & 1 & f \\ 0 & 0 & 0 & 1 \end{bmatrix} \cdot \begin{bmatrix} 1 & 0 & 0 & 0 \\ 0 & 0 & -1 & 0 \\ 0 & 1 & 0 & 0 \\ 0 & 0 & 0 & 1 \end{bmatrix} \cdot \underbrace{\begin{bmatrix} 0 & -1 & 0 & 0 \\ 1 & 0 & 0 & 0 \\ 0 & 0 & 1 & 0 \\ 0 & 0 & 0 & 1 \end{bmatrix} \begin{bmatrix} a \\ b \\ c \\ 1 \end{bmatrix}}$$

(6.13)

$$= \underbrace{\begin{bmatrix} 1 & 0 & 0 & d \\ 0 & 0 & -1 & e \\ 0 & 1 & 0 & f \\ 0 & 0 & 0 & 1 \end{bmatrix}} \qquad \underbrace{\begin{bmatrix} -b \\ a \\ c \\ 1 \end{bmatrix}}$$

$$= \begin{bmatrix} -b + d \\ -c + e \\ a + f \\ 1 \end{bmatrix}$$

In Example 6.2, we made use of the fact that matrix multiplication is associative by combining the first two matrices into a single matrix by multiplying them together. This can be done in general, allowing us to define a single matrix which is a *homogeneous transform*. Again, using the previous example, the transform matrix is:

$$[\text{Trans}(d, e, f)] \cdot [\text{Rot}(x, 90)] \cdot [\text{Rot}(z, 90)]$$

$$= \begin{bmatrix} 1 & 0 & 0 & d \\ 0 & 1 & 0 & e \\ 0 & 0 & 1 & f \\ 0 & 0 & 0 & 1 \end{bmatrix} \cdot \begin{bmatrix} 1 & 0 & 0 & 0 \\ 0 & 0 & -1 & 0 \\ 0 & 1 & 0 & 0 \\ 0 & 0 & 0 & 1 \end{bmatrix} \cdot \begin{bmatrix} 0 & -1 & 0 & 0 \\ 1 & 0 & 0 & 0 \\ 0 & 0 & 1 & 0 \\ 0 & 0 & 0 & 1 \end{bmatrix}$$

$$= \begin{bmatrix} 0 & -1 & 0 & d \\ 0 & 0 & -1 & e \\ 1 & 0 & 0 & f \\ 0 & 0 & 0 & 1 \end{bmatrix}$$

(6.14)

We must emphasize the right-to-left nature of the operations in this example. The rotation about $z$ occurs *before* the rotation about $x$, which occurs *before* the translation. We also remind the reader that matrix multiplication is not commutative and that neither are rotations. Specifically, $\text{Rot}(x, \theta) \, \text{Rot}(y, \phi) \neq \text{Rot}(y, \phi) \, \text{Rot}(x, \theta)$ in general.

### The General Form of the Transformation Matrix

In Section 6.1, we defined roll, pitch, and yaw intuitively as motions of the wrist. We will now be somewhat more rigorous and specify them as rotations about coordinate axes. Specifically,

$$\text{Roll}(\phi_z) = \text{rotate } \phi_z \text{ about } z$$

$$\text{Pitch}(\phi_y) = \text{rotate } \phi_y \text{ about } y$$

$$\text{Yaw}(\phi_x) = \text{rotate } \phi_x \text{ about } x$$

Thus, we may define a combination of these as

$$\text{RPY}(\phi_z, \phi_y, \phi_x) = \text{Rot}(z, \phi_z) \, \text{Rot}(y, \phi_y) \, \text{Rot}(x, \phi_x)$$

In expanding this, we abbreviate cos by $c$ and sin by $s$, in order to write the result compactly:

$$\text{RPY}(\phi_z, \phi_y, \phi_x)$$

$$= \begin{bmatrix} c\phi_z c\phi_y & c\phi_z s\phi_y s\phi_x - s\phi_z c\phi_x & c\phi_z s\phi_y c\phi_x + s\phi_z s\phi_x & 0 \\ s\phi_z c\phi_y & s\phi_z s\phi_y s\phi_x + c\phi_z c\phi_x & s\phi_z s\phi_y c\phi_x - c\phi_z s\phi_x & 0 \\ -s\phi_y & c\phi_y s\phi_x & c\phi_y c\phi_x & 0 \\ 0 & 0 & 0 & 1 \end{bmatrix}$$

$$(6.15)$$

Finally, we may, if we wish, replace the zeros in the fourth column with $a$, $b$, $c$ to include the effects of translation and, therefore, have a complete transformation matrix, including three rotations and three translations.

## 6.5 COORDINATE FRAMES

We will often use transformation matrices to describe the location of one coordinate system relative to another. When used in this way, we will refer to such a system as a *coordinate frame*. Use of a transformation to describe a coordinate system is meaningful *only* when it is stated as relative to another frame. With such an interpretation, the origin of the second frame may be found as the transformation of the point $[0 \ 0 \ 0 \ 1]^{\mathsf{T}}$. A unit vector in the $x$ direction in the new system is the transformation of the point $[1 \ 0 \ 0 \ 1]^T$. Similarly, unit vectors in the $y$ and $z$ directions of the new system are transformations of $[0 \ 1 \ 0 \ 1]^T$ and $[0 \ 0 \ 1 \ 1]^T$, respectively.

### Example 6.3 Coordinate Frames

Find the origin and coordinate directions of a frame resulting from a rotation of $90°$ about the $z$ axis, followed by a displacement of $[1, 7, 3]^T$, as shown in Figure 6.5.

**Solution:** The origin of the new system is found by transforming $[0 \ 0 \ 0 \ 1]^T$.

$$
\begin{bmatrix} 1 \\ 7 \\ 3 \\ 1 \end{bmatrix} = \begin{bmatrix} 0 & -1 & 0 & 1 \\ 1 & 0 & 0 & 7 \\ 0 & 0 & 1 & 3 \\ 0 & 0 & 0 & 1 \end{bmatrix} \cdot \begin{bmatrix} 0 \\ 0 \\ 0 \\ 1 \end{bmatrix} \tag{6.16}
$$

The origin of the new coordinate system is thus at $[1, 7, 3]^T$ in the old system. A unit vector in the $x$ direction in the new system is

$$
\begin{bmatrix} 1 \\ 8 \\ 3 \\ 1 \end{bmatrix} = \begin{bmatrix} 0 & -1 & 0 & 1 \\ 1 & 0 & 0 & 7 \\ 0 & 0 & 1 & 3 \\ 0 & 0 & 0 & 1 \end{bmatrix} \cdot \begin{bmatrix} 1 \\ 0 \\ 0 \\ 1 \end{bmatrix} \tag{6.17}
$$

Similarly, we find that a unit vector in the $y$ direction transforms to $[0, 7, 3, 1]$ and that a unit vector in the $z$ direction is transformed to $[1, 7, 4, 1]$.

In this example the coordinate frame is defined by the transform consisting of a rotation of $90°$ about the $z$ axis. The locations of the transformed unit vectors are indicated by stars, and their directions with respect to the transformed origin define the transformed coordinate directions.

Let us assume that the original frame in Figure 6.5 is named $R$. (We will often use this term to represent the coordinate system whose origin is at the base of the manipulator.) Now, given the point $P = [2, 2, 0, 1]^T$ in frame $H$ (denoted $^H P$), we can find $^R P$ by $^R P = {}^R T_H \cdot {}^H P$, where $^R T_H$ is the transformation matrix, just given, which relates frames $H$ and $R$.

$$
R_P = \begin{bmatrix} 0 & -1 & 0 & 1 \\ 1 & 0 & 0 & 7 \\ 0 & 0 & 1 & 3 \\ 0 & 0 & 0 & 1 \end{bmatrix} \cdot \begin{bmatrix} 2 \\ 2 \\ 0 \\ 1 \end{bmatrix} = \begin{bmatrix} -1 \\ 9 \\ 3 \\ 1 \end{bmatrix} \tag{6.18}
$$

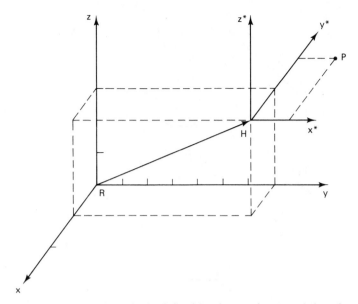

**Figure 6.5** The coordinate frame defined by the transform consisting of a rotation of 90° about the $z$ axis. The locations of the transformed unit vectors are indicated by stars, and their directions with respect to the transformed origin define the transformed coordinate directions.

We can now see how the concept of coordinate frames relates to robotics. $^{H}P$ defines a point on the hand of the robot (perhaps the tip of a tool) which is known relative to the hand frame $H$. The location of the hand is known relative to the base frame $R$, by the transformation matrix $^{R}T_{H}$. Thus we are able to find the location of the tool tip in base coordinates.

We extend this concept somewhat further now to show its full usefulness with another example.

Consider the situation shown in Figure 6.6. The robot is holding a drill and drilling a hole in a part. This is the *desired* state; we must solve a problem to find the conditions which will make it true. The base of the robot is the origin of a frame named $R$ whose location is known relative to a *universe frame $U$*. That knowledge is embedded in the transformation $^{U}T_{R}$, assumed to be constant. When in the proper location, the hand position will be related to $R$ by a transform $^{R}T_{H}$. This is initially unknown, and we must solve for it. The tip of the tool is related to $H$ by a transform $^{H}T_{E}$, assumed constant (if the tool doesn't slip in the hand). Thus the location of the drill bit may be related to the universe frame by

$$^{U}T_{E} = {^{U}T_{R}} \cdot {^{R}T_{H}} \cdot {^{H}T_{E}}$$

Futhermore, the location of the hole to be drilled on the part may be related by $U$ by

$$^{U}T_{E} = {^{U}T_{P}} \cdot {^{P}T_{E}}$$

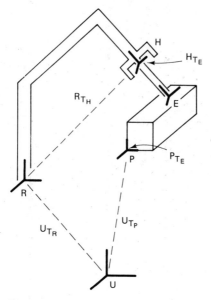

**Figure 6.6**  Important coordinate frames in a robot application: $U$, universe frame; $R$, robot base; $H$, hand; $E$, tool tip; $P$, workpiece.

and we have two descriptions for the same point. Equating them, we have

$$^U T_R \cdot {}^R T_H \cdot {}^H T_E = {}^U T_P \cdot {}^P T_E \tag{6.19}$$

In virtually all robot applications, it is $^R T_H$ that must be found. It is that transform that specifies the location of the hand with respect to the base. To solve for $^R T_H$ in the preceding equation, we first postmultiply by $^H T_E^{-1}$.

$$^U T_R \cdot {}^R T_H = {}^U T_P \cdot {}^P T_E \cdot {}^H T_E^{-1} \tag{6.20}$$

and premultiply by $^U T_R^{-1}$ and find that

$$^R T_H = {}^U T_R^{-1} \cdot {}^U T_P \cdot {}^P T_E \cdot {}^H T_E^{-1} \tag{6.21}$$

### Inverting Homogeneous Transforms

Let us assign names to the elements of a homogeneous transform as

$$T = \begin{bmatrix} n_x & o_x & a_x & p_x \\ n_y & o_y & a_y & p_y \\ n_z & o_z & a_z & p_z \\ 0 & 0 & 0 & 1 \end{bmatrix} \tag{6.22}$$

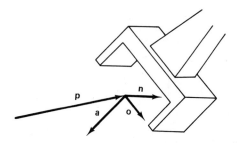

**Figure 6.7** Definition of approach, **a**, orientation, **o**, and normal, **n**, vectors in the hand frame; **p** determines the location of the frame, and any two of **n**, **o**, and **a** are sufficient to specify its orientation.

The columns of the hand frame, $^R T_H$, define three orthogonal vectors which together specify the hand orientation (see Figure 6.7). These are referred to as the *approach vector*, which points in the direction of the wrist; the *orientation vector*, which, together with the approach vector, specifies the orientation of the hand; and the *normal vector*, which is orthogonal to the other two. Hence the use of $a$, $o$, and $n$ in the definition of the hand frame. In future sections, we will use these same names for the columns of transform matrices, even though we may not be specifically referring to the hand frame.

It can be shown that

$$T^{-1} = \begin{bmatrix} n_x & n_y & n_z & -\mathbf{p} \cdot \mathbf{n} \\ o_x & o_y & o_z & -\mathbf{p} \cdot \mathbf{o} \\ a_x & a_y & a_z & -\mathbf{p} \cdot \mathbf{a} \\ 0 & 0 & 0 & 1 \end{bmatrix} \qquad (6.23)$$

where $\mathbf{p} \cdot \mathbf{n}$ represents the dot product of the vectors $\mathbf{p}$ and $\mathbf{n}$. That is,

$$\mathbf{p} \cdot \mathbf{n} = p_x n_x + p_y n_y + p_z n_z \qquad (6.24a)$$

Similarly,

$$\mathbf{p} \cdot \mathbf{o} = p_x o_x + p_y o_y + p_z o_z \qquad (6.24b)$$

and

$$\mathbf{p} \cdot \mathbf{a} = p_x a_x + p_y a_y + p_z a_z \qquad (6.24c)$$

**Example 6.4  Inverting Homogeneous Transforms**

Find the inverse of the following transformation matrix. Prove it is the inverse.

$$H = \begin{bmatrix} 0 & 0 & 1 & 1 \\ 0 & 1 & 0 & 2 \\ -1 & 0 & 0 & 3 \\ 0 & 0 & 0 & 1 \end{bmatrix}$$

**Solution:**   Applying Eq. 6.23, we find

$$H^{-1} = \begin{bmatrix} 0 & 0 & -1 & 3 \\ 0 & 1 & 0 & -2 \\ 1 & 0 & 0 & -1 \\ 0 & 0 & 0 & 1 \end{bmatrix}$$

We prove this is the inverse by multiplying it by the original matrix.

$$\begin{bmatrix} 0 & 0 & 1 & 1 \\ 0 & 1 & 0 & 2 \\ -1 & 0 & 0 & 3 \\ 0 & 0 & 0 & 1 \end{bmatrix} \cdot \begin{bmatrix} 0 & 0 & -1 & 3 \\ 0 & 1 & 0 & -2 \\ 1 & 0 & 0 & -1 \\ 0 & 0 & 0 & 1 \end{bmatrix}$$

$$= \begin{bmatrix} 1 & 0 & 0 & 0 \\ 0 & 1 & 0 & 0 \\ 0 & 0 & 1 & 0 \\ 0 & 0 & 0 & 1 \end{bmatrix}$$

This method of matrix inversion is much faster than is the general method; however, it is not applicable to arbitrary 4 × 4 matrices but results from the peculiar nature of homogeneous transformations.

One other useful property of homogeneous transforms can be derived. That is,

$$\mathbf{n} \times \mathbf{o} = \mathbf{a} \qquad (6.25a)$$

$$\mathbf{a} \times \mathbf{n} = \mathbf{o} \qquad (6.25b)$$

$$\mathbf{o} \times \mathbf{a} = \mathbf{n} \qquad (6.25c)$$

Here, × represents the cross product of two vectors. $\mathbf{n} \times \mathbf{o} = \mathbf{a}$ may be expanded to

$$a_x = n_y o_z - o_y n_z \qquad (6.26a)$$

$$a_y = o_x n_z - n_x o_z \qquad (6.26b)$$

$$a_z = n_x o_y - n_y o_x \qquad (6.26c)$$

The other two equations may be similarly expanded.

This fact allows us to save some computer time later, since it defines one column of the $T$ matrix in terms of the others.

$^R T_H$ tells us how to position the robot hand so that it will be properly located to drill the hole as desired. This 4 × 4 matrix relates only the position and orientation of the hand in Cartesian space to the position and orientation of the robot base, again in Cartesian space. It does not tell us how to move the various joints of the robot to achieve that hand location. To accomplish that, we must study the relationship between joint angles and hand location. These relationships are the subject of the next chapter.

## 6.6 SYNOPSIS

### Vocabulary

You should know the definition and application of the following terms.

approach vector
Cartesian space
coordinate frames
configuration
degree of freedom
frame
homogeneous coordinates
homogeneous transform
normal vector
orientation
orientation vector
pitch
position vector
roll
rotation
translation
universe frame
yaw

### Notation

| Symbol | Meaning |
| --- | --- |
| $\theta$ | An arbitrary angle |
| $s\theta$ | Shorthand for sin $(\theta)$ |
| $c\theta$ | Shorthand for cos $(\theta)$ |

| Symbol | Meaning |
|--------|---------|
| $\text{Rot}(x, \theta)$ | A 4 × 4 transform matrix representing a rotation of $\theta$ about the $x$ axis |
| $\text{Trans}(d, e, f)$ | A 4 × 4 transform matrix representing a translation of $d$ units on the $x$ axis, $e$ units on $y$, and $f$ units on $z$ |
| $\phi_z$ | Rotation about the $z$ axis of the hand frame (roll) |
| $\phi_y$ | Rotation about the $y$ axis of the hand frame (pitch) |
| $\phi_x$ | Rotation about the $x$ axis of the hand frame (yaw) |
| $\text{RPY}(\phi_z, \phi_y, \phi_x)$ | A 4 × 4 transform matrix representing the effect of $\phi_z$ radians of roll, $\phi_y$ radians of pitch, and $\phi_x$ radians of yaw. |
| $H$ | The hand frame |
| $U$ | The universe frame (often chosen equal to $R$) |
| $R$ | The robot base frame |
| $\mathbf{n}$ | The normal vector (first column of $^R T_H$) |
| $\mathbf{o}$ | The orientation vector (second column of $^R T_H$) |
| $\mathbf{a}$ | The approach vector (third column of $^R T_H$) |
| $\mathbf{p}$ | The position vector (fourth column of $^R T_H$) |

## 6.7 PROBLEMS

1. Write the homogeneous transform matrix that represents
   (a) a rotation of 90° about the $z$ axis, followed by
   (b) a rotation of -90° about the $x$ axis, followed by
   (c) a translation of $\langle 3, 7, 9 \rangle$

2. Write the homogeneous transform matrix that represents
   (a) a translation of $\langle 3, 7, 9 \rangle$, followed by
   (b) a rotation of 90° about the $z$ axis, followed by
   (c) a rotation of -90° about the $x$ axis

3. The matrix

$$\begin{bmatrix} ? & 0 & -1 & 0 \\ ? & 0 & 0 & 1 \\ ? & -1 & 0 & 2 \\ ? & 0 & 0 & 1 \end{bmatrix}$$

is known to represent a homogeneous coordinate transform. Find the elements designated by the question marks. Use Eq. 6.25.

4. Use Eq. 6.23 to find the inverse of the matrix

$$
\begin{bmatrix}
0 & 1 & 0 & -1 \\
0 & 0 & -1 & 2 \\
-1 & 0 & 0 & 0 \\
0 & 0 & 0 & 1
\end{bmatrix}
$$

5. A part is located relative to the universe frame by the transform

$$
^{U}T_{P} =
\begin{bmatrix}
0 & 1 & 0 & -1 \\
0 & 0 & -1 & 2 \\
-1 & 0 & 0 & 0 \\
0 & 0 & 0 & 1
\end{bmatrix}
$$

A robot base is located relative to the universe frame by

$$
^{U}T_{R} =
\begin{bmatrix}
1 & 0 & 0 & 1 \\
0 & 1 & 0 & 5 \\
0 & 0 & 1 & 9 \\
0 & 0 & 0 & 1
\end{bmatrix}
$$

We wish to put the hand on the part. That is, we wish to align the hand frame and the part frame. What is the transformation $^{R}T_{H}$ that makes this happen?

6. Use the matrices given in problem 5, consider only the *position* of the hand.
   (a) What is the position of the part relative to the universe frame?
   (b) What is the position of the robot relative to the universe frame?
   (c) Perform vector addition or subtraction (which is correct?) to find the part relative to the robot.
   (d) Now look at column 4 of your solution to problem 5. Did you work problem 5 correctly?

7. Rotational velocity and rotational acceleration are vector quantities. Is rotational displacement also a vector quantity?

For problems 8 to 11, consider the following (refer to Figure 6.6). Let

$$
^{U}T_{R} =
\begin{bmatrix}
0 & 1 & 0 & 2 \\
-1 & 0 & 0 & 4 \\
0 & 0 & 1 & 5 \\
0 & 0 & 0 & 1
\end{bmatrix}
$$

$$^R T_H = \begin{bmatrix} 1 & 0 & 0 & 1 \\ 0 & -1 & 0 & 3 \\ 0 & 0 & -1 & 2 \\ 0 & 0 & 0 & 1 \end{bmatrix}$$

8. (a) Explain how $R$ is oriented relative to $U$ (i.e., what type of rotations or translations have been implemented to derive frame $R$ from frame $U$). (b) Repeat the above for frame $H$ relative to $R$. Assume that only one rotation has been implemented in both cases.

9. (a) Given a vector $[1, 2, 3]$ in frame $H$, find its expression in frame $R$. (b) Find the expression of vector $[1, 2, 3]$ (in frame $H$) in frame $U$.

10. Find $^U T_H$ and verify the result obtained in 8(b).

11. Given $^U T_H$ as earlier, find $^U T_H^{-1}$.

# 7

# KINEMATICS OF POSITION

The term *kinematics* is defined by Sandor (1983) as follows:

Kinematics deals with the geometry and the time-dependent aspects of motion without considering the forces causing motion. In these studies, forces may or may not be associated with the motion. The parameters of interest in kinematics are position, displacement, velocity, acceleration, and time.

In the previous chapter, we learned how to relate one coordinate frame to another through the use of 4 X 4 matrices called homogeneous transformations. We also learned that the transformation $^R T_H$, which relates the hand frame to the robot base frame, is the critical transformation, for it tells us how to position the hand. Unfortunately, $^R T_H$ does not tell us how to move the various joints of the robot to achieve this hand configuration. The subject of kinematics is used to relate the hand position to the joint variables. In this chapter, we will learn how to accomplish this for arbitrary arm geometries.

## 7.1 RELATIONS BETWEEN LINKS AND JOINTS

A manipulator can, in general, be described as a series of links connected at joints. The angles between the links, called the *joint angles,* are typical joint variables. However, some types of links (prismatic joints) can grow longer, and in that case the joint variable may actually be the length of the link.

The position and orientation of one link is related to the next (adjacent) link by a homogeneous transform, known as an $A$ matrix. $A_1$ relates the first link to the manipulator base, $A_2$ relates the second link to the first link, and so on. Hence, the configuration of any link may be found by multiplying by the appropriate number of $A$ matrices. In particular,

$$^R T_H = A_1 \cdot A_2 \cdot A_3 \cdot A_4 \cdot A_5 \cdot A_6 \qquad (7.1)$$

To be able to specify the $A$ matrices, we need to examine closely the geometry of links.

In this definition of link coordinate systems, we will follow the notation of Paul (1981). To make the explanation clearer, we will ignore the possibility of bent links. That is, we assume that a link is a straight rod terminating in a rotary joint, with the rotary joints at both ends having parallel or perpendicular axes (Figure 7.1). In the case of parallel axes, we allow the axis of the second joint to be rotated in a plane perpendicular to the link, through a *twist angle* $\alpha$. The notation developed by Paul (1981) does not require either that the link be straight or that the twist be in a perpendicular plane and, hence, is more general. However, the notation developed here is adequate for virtually all commercial robots.

### 7.1.1 Kinematic Equations of the $\theta$-$r$ Manipulator

In earlier chapters, we defined a manipulator with only two actuators, one prismatic, $r$, and one rotary, $\theta$. With only two degrees of freedom, motion for this manipulator is restricted to the $x$-$y$ plane, and no control of orientation is possible. The *forward kinematic transform* for this manipulator relates the $x$-$y$ position of the hand to the joint variables as shown in Example 7.1.

**Example 7.1  Kinematics of the $\theta$-$r$ Manipulator**

In deriving this transform, we could use geometric intuition and recognize the immediate relationship among $r$, $\theta$, $x$, and $y$. Rather than take that approach, we will be more rigorous and will define two coordinate frames, one for each joint, and relate them. In this way, we will develop a simple example of use of the $A$ matrices which will help us in developing the transforms for more complex, realistic manipulators.

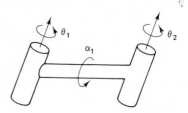

**Figure 7.1**  Two rotational joints.

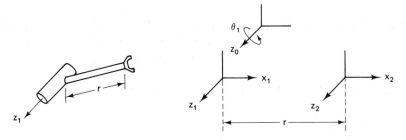

**Figure 7.2**   The $\theta$-$r$ manipulator.

We consider the $\theta$-$r$ manipulator as composed of two links as shown in Figure 7.2. The first link has a length of zero, and the second is a prismatic joint whose length is $r$, the joint variable.

The origin of frame 2 is related to the origin of frame 1 by a simple translation, represented by the $A$ matrix,

$$A_2 = \begin{bmatrix} 1 & 0 & 0 & r \\ 0 & 1 & 0 & 0 \\ 0 & 0 & 1 & 0 \\ 0 & 0 & 0 & 1 \end{bmatrix} \tag{7.2}$$

Similarly, frame 1 is related to base coordinates (frame 0) by a homogeneous transform having only rotation about the $z$ axis:

$$A_1 = \begin{bmatrix} \cos\theta & -\sin\theta & 0 & 0 \\ \sin\theta & \cos\theta & 0 & 0 \\ 0 & 0 & 1 & 0 \\ 0 & 0 & 0 & 1 \end{bmatrix} \tag{7.3}$$

The following transform then results:

$$^R T_H = \begin{bmatrix} \cos\theta & -\sin\theta & 0 & 0 \\ \sin\theta & \cos\theta & 0 & 0 \\ 0 & 0 & 1 & 0 \\ 0 & 0 & 0 & 1 \end{bmatrix} \begin{bmatrix} 1 & 0 & 0 & r \\ 0 & 1 & 0 & 0 \\ 0 & 0 & 1 & 0 \\ 0 & 0 & 0 & 1 \end{bmatrix} \tag{7.4}$$

$$= \begin{bmatrix} \cos\theta & -\sin\theta & 0 & r\cos\theta \\ \sin\theta & \cos\theta & 0 & r\sin\theta \\ 0 & 0 & 1 & 0 \\ 0 & 0 & 0 & 1 \end{bmatrix} \tag{7.5}$$

This transform then relates a new coordinate system located in the hand to the base coordinates. We can see that this agrees with intuition by noting that the origin of this new frame is at the transform of $[0 \ 0 \ 0 \ 1]^T$, which is

$$[r\cos\theta, r\sin\theta, 0, 1]$$

This manipulator provides an excellent example of the distinction between "rotate and translate" and "translate and rotate." The general

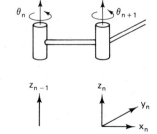

**Figure 7.3**  Two rotational joints with parallel axes.

homogeneous transform is an example of the former and the $\theta$-$r$ manipulator an example of the latter. The student should pay close attention to the fourth column of the $T$ matrix and convince himself or herself of the difference between these two cases.

The choice of coordinate systems used to relate one joint or one link to another is totally arbitrary. For example, in Figure 7.2 we chose coordinate systems for joints 1 and 2 that had parallel $z$ axes. In this way, the results provided a simple example which agreed with intuition. In the more general case, however, parallel axes are seldom feasible or desirable. In the next section, we will discuss a more general technique for assigning coordinate systems to joints.

### 7.1.2 Coordinate Frame Definition: for Rotary Joints (see Figure 7.3)

1. $Z_n$ is chosen coincident with the axis of joint $n + 1$.
2. $X_n$ is chosen coincident with the link, pointing away from the link.

In this case, we have

1. A rotation of $\alpha_n$ about the link ($x$ axis)
2. A translation of $d_n$ along the $x$ axis
3. A rotation of $\theta_n$ about $Z_{n-1}$

$$
A = \begin{bmatrix} \cos\theta & -\sin\theta & 0 & 0 \\ \sin\theta & \cos\theta & 0 & 0 \\ 0 & 0 & 1 & 0 \\ 0 & 0 & 0 & 1 \end{bmatrix} \begin{bmatrix} 1 & 0 & 0 & d \\ 0 & 1 & 0 & 0 \\ 0 & 0 & 1 & 0 \\ 0 & 0 & 0 & 1 \end{bmatrix}
$$

$$
\times \begin{bmatrix} 1 & 0 & 0 & 0 \\ 0 & \cos\alpha & -\sin\alpha & 0 \\ 0 & \sin\alpha & \cos\alpha & 0 \\ 0 & 0 & 0 & 1 \end{bmatrix}
$$

$$(7.6)$$

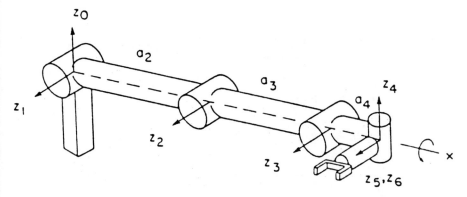

**Figure 7.4**    (Paul) An articulated manipulator. (*Courtesy MIT Press*)

To simplify the notation, we will abbreviate $\cos\theta$ and $\sin\theta$, by $c\theta$ and $s\theta$, and so on.* Multiplying the transformations, we find the generalized $A$ matrix for a rotary joint.

$$A = \begin{bmatrix} c\theta & -s\theta c\alpha & s\theta s\alpha & dc\theta \\ s\theta & c\theta c\alpha & -c\theta s\alpha & ds\theta \\ 0 & s\alpha & c\alpha & 0 \\ 0 & 0 & 0 & 1 \end{bmatrix} \qquad (7.7)$$

### Example 7.2  Coordinate Frame Definitions

Figure 7.1 shows two rotary joints connected by a link. Suppose that these are the first two links of the articulated manipulator shown in Figure 7.4. Determine $A_1$.

**Solution:**    The center of rotation for joint 1 is about the $Z_0$ axis. Furthermore, the origin of coordinate system 1 is at the center of rotation for joint 2. Since these two centers of rotation intersect, we have $d_1 = 0$. However, the $Z_1$ axis is rotated by 90° relative to joint 0, so $\alpha = 90°$. Then

$$A_1 = \begin{bmatrix} c_1 & 0 & s_1 & 0 \\ s_1 & 0 & -c_1 & 0 \\ 0 & 1 & 0 & 0 \\ 0 & 0 & 0 & 1 \end{bmatrix}$$

Figure 7.5 illustrates the assignment of coordinate systems to the links of an articulated manipulator (Figure 7.4). Use of these coordinate system assignments in conjunction with Eq. 7.6 results in the $A$ matrices shown in Figure 7.6.

---

*When we are confronted with a number of joint variables, $\theta_1$, $\theta_2$, ..., and so on, we will further abbreviate the notation by using $s_1 = \sin(\theta_1)$, $c_1 = \cos(\theta_1)$, and so on.

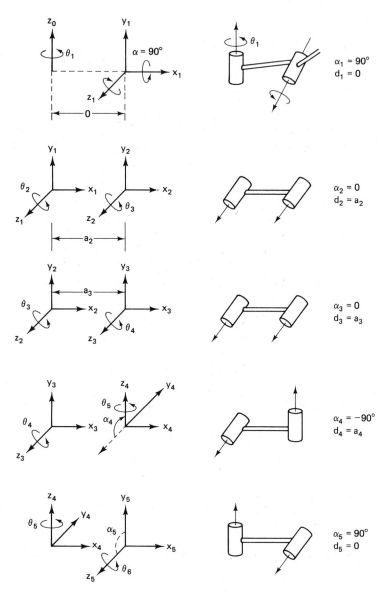

**Figure 7.5**   Pairs of joints of an articulated manipulator.   Note that some links have a length of zero.

### 7.1.3 Coordinate Frame Definitions: Prismatic Joints

If a robot has one or more prismatic joints, the mechanism for deriving the $A$ matrices is different, since the links can grow longer. To avoid confusing the student with too many variables at once, we will

$$A_1 = \begin{bmatrix} C_1 & 0 & S_1 & 0 \\ S_1 & 0 & -C_1 & 0 \\ 0 & 1 & 0 & 0 \\ 0 & 0 & 0 & 1 \end{bmatrix} \qquad A_4 = \begin{bmatrix} C_4 & 0 & -S_4 & C_4 a_4 \\ S_4 & 0 & C_4 & S_4 a_4 \\ 0 & -1 & 0 & 0 \\ 0 & 0 & 0 & 1 \end{bmatrix}$$

$$A_2 = \begin{bmatrix} C_2 & -S_2 & 0 & C_2 a_2 \\ S_2 & C_2 & 0 & S_2 a_2 \\ 0 & 0 & 1 & 0 \\ 0 & 0 & 0 & 1 \end{bmatrix} \qquad A_5 = \begin{bmatrix} C_5 & 0 & S_5 & 0 \\ S_5 & 0 & -C_5 & 0 \\ 0 & 1 & 0 & 0 \\ 0 & 0 & 0 & 1 \end{bmatrix}$$

$$A_3 = \begin{bmatrix} C_3 & -S_3 & 0 & C_3 a_3 \\ S_3 & C_3 & 0 & S_3 a_3 \\ 0 & 0 & 1 & 0 \\ 0 & 0 & 0 & 1 \end{bmatrix} \qquad A_6 = \begin{bmatrix} C_6 & -S_6 & 0 & 0 \\ S_6 & C_6 & 0 & 0 \\ 0 & 0 & 1 & 0 \\ 0 & 0 & 0 & 1 \end{bmatrix}$$

**Figure 7.6**    (Paul) $A$ matrices for the articulated manipulator.

deal here only with articulated manipulators. An example deriving $A$ matrices for a robot having a prismatic joint is given in the appendix to this chapter.

### 7.1.4 Computing $^R T_H$ from the $A$ Matrices

By multiplying the $A$ matrices together, as shown in Eq. 7.1, we can determine a simple transform that relates the position and orientation of the hand to the robot base, a transformation which is a function of the joint angles.

**Example 7.3 Computing $^R T_H$**

Figure 7.6 describes the $A$ matrices for a six-degree-of-freedom robot. Suppose that this robot has no wrist. With only three joints ($\theta_1$, $\theta_2$, $\theta_3$), it should still be possible to control the *position* of the hand, although not the orientation. Derive $^R T_H$ using only $A_1$, $A_2$, and $A_3$. Then, show the relationship between hand position, expressed in $x, y, z$, and joint angles.

**Solution:**    Multiplying the $A$ matrices, we have, for this three-jointed manipulator,

$$^R T_H = A_1 \cdot A_2 \cdot A_3$$

$$\begin{bmatrix} c_1 & 0 & s_1 & 0 \\ s_1 & 0 & -c_1 & 0 \\ 0 & 1 & 0 & 0 \\ 0 & 0 & 0 & 1 \end{bmatrix} \begin{bmatrix} c_2 & -s_2 & 0 & c_2 a_2 \\ s_2 & c_2 & 0 & s_2 a_2 \\ 0 & 0 & 1 & 0 \\ 0 & 0 & 0 & 1 \end{bmatrix} \begin{bmatrix} c_3 & -s_3 & 0 & c_3 a_3 \\ s_3 & c_3 & 0 & s_3 a_3 \\ 0 & 0 & 1 & 0 \\ 0 & 0 & 0 & 1 \end{bmatrix}$$

$$= \begin{bmatrix} c_1 & 0 & s_1 & 0 \\ s_1 & 0 & -c_1 & 0 \\ 0 & 1 & 0 & 0 \\ 0 & 0 & 0 & 1 \end{bmatrix} \times$$

$$
\begin{bmatrix}
c_2c_3 - s_2s_3 & -c_2s_3 - s_2c_3 & 0 & c_2c_3a_3 - s_2c_3a_3 + c_2a_2 \\
s_2c_3 - s_3c_2 & -s_2s_3 + c_2c_3 & 0 & s_2c_3a_3 + c_2s_3a_3 + s_2a_2 \\
0 & 0 & 1 & 0 \\
0 & 0 & 0 & 1
\end{bmatrix}
$$

$$
= \begin{bmatrix}
c_1c_2c_3 - c_1s_2s_3 & -c_1c_2s_3 - c_1s_2c_3 & s_1 & c_1c_2c_3a_3 - c_1s_2c_3a_3 + c_1c_2a_2 \\
s_1c_2c_3 - s_1s_2s_3 & -s_1c_2s_3 - s_1s_2c_3 & -c_1 & s_1c_2c_3a_3 - s_1s_2c_3a_3 + s_1c_2a_2 \\
s_2c_3 - s_3c_2 & -s_2s_3 + c_2c_3 & 0 & s_2c_3a_3 + c_2s_3a_3 + s_2a_2 \\
0 & 0 & 0 & 1
\end{bmatrix}
$$

We can determine the position of the hand from the fourth column of $^R T_H$:

$$
x = c_1c_2c_3a_3 - c_1s_2c_3a_3 + c_1c_2a_2
$$

$$
y = s_1c_2c_3a_3 - s_1s_2c_3a_3 + s_1c_2a_2
$$

$$
z = s_2c_3a_3 + c_2s_3a_3 + s_2a_2
$$

### 7.1.5 Kinematic Equations for the Articulated Manipulator

We define

$$
^R T_H = \begin{bmatrix}
n_x & o_x & a_x & p_x \\
n_y & o_y & a_y & p_y \\
n_z & o_z & a_z & p_z \\
0 & 0 & 0 & 1
\end{bmatrix} \tag{7.8}
$$

where

$$
\begin{aligned}
n_x &= c_1[c_{234}c_5c_6 - s_{234}s_6] - s_1s_5c_6 \\
n_y &= s_1[c_{234}c_5c_6 - s_{234}s_6] + c_1s_5c_6 \\
n_z &= s_{234}c_5c_6 + c_{234}s_6
\end{aligned}
$$

$$
\begin{aligned}
o_x &= -c_1[c_{234}c_5s_6 + s_{234}c_6] + s_1s_5s_6 \\
o_y &= -s_1[c_{234}c_5s_6 + s_{234}c_6] - c_1s_5s_6 \\
o_z &= -s_{234}c_5s_6 + c_{234}c_6 \\
a_x &= c_1c_{234}s_5 + s_1c_5 \\
a_y &= s_1c_{234}s_5 - c_1c_5 \\
a_z &= s_{234}s_5
\end{aligned} \tag{7.9}
$$

$$
\begin{aligned}
p_x &= c_1[c_{234}a_4 + c_{23}a_3 + c_2a_2] \\
p_y &= s_1[c_{234}a_4 + c_{23}a_3 + c_2a_2] \\
p_z &= s_{234}a_4 + s_{23}a_3 + s_2a_2
\end{aligned}
$$

Note: This solution contains redundant calculations. From Eq. 6.25, we know that only two of **n**, **o**, and **a** need to be evaluated explicitly, since the other can be obtained by six multiplies and three adds.

The matrix $^R T_H$ provides a relationship between the position and orientation of the hand and a coordinate system located at the robot base. Specifically, if we know joint variables $(\theta_1 - \theta_6)$, then substitution into Eq. 7.8 will produce a homogeneous transform specifying the position and orientation of the hand.

**Example 7.4  Evaluation of Kinematic Equations**

For the following set of values,

$$
\begin{aligned}
\theta_1 &= 90° \\
\theta_2 &= 0° \\
\theta_3 &= 60° \\
\theta_4 &= 90° \\
\theta_5 &= 0° \\
\theta_6 &= 30°
\end{aligned}
$$

show, numerically, that $\mathbf{a} = \mathbf{n} \times \mathbf{o}$.

**Solution:**    We first evaluate $\mathbf{n}$, $\mathbf{o}$, and $\mathbf{a}$ for these joint angles.

$$
\begin{array}{lll}
n_x = 0 & o_x = 0 & a_x = 1 \\
n_y = -1 & o_y = 0 & a_y = 0 \\
n_z = 0 & o_z = -1 & a_z = 0
\end{array}
$$

Then from Eq. 6.26,

$$
a_x = n_y o_z - o_y n_z = (-1)(-1) - (0)(0) = 1
$$

$$
a_y = o_x n_z - n_x o_z = (0)(0) - (0)(-1) = 0
$$

$$
a_z = n_x o_y - n_y o_x = (0)(0) - (-1)(0) = 0
$$

## 7.2 DETERMINING ORIENTATION: AN INVERSE PROBLEM

As was discussed in Chapter 6, orientation of the fingers can be defined in several ways. These include roll, pitch, and yaw (RPY) and Euler angles. In the previous section, we learned how to specify $^R T_H$, a homogeneous transformation matrix in which both the position and orientation of the fingers is represented. This matrix may be written as

$$
T = \begin{bmatrix}
n_x & o_x & a_x & p_x \\
n_y & o_y & a_y & p_y \\
n_z & o_z & a_z & p_z \\
0 & 0 & 0 & 1
\end{bmatrix}
\tag{7.10}
$$

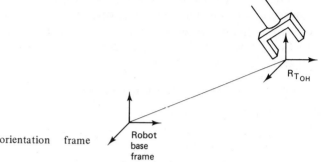

Figure 7.7 Hand orientation frame $T_{OH}$.

Since we are only interested in the orientation of the fingers, we will move the coordinate system from the base of the robot by a translation:

$$
T_{OH} = \begin{bmatrix} 1 & 0 & 0 & -p_x \\ 0 & 1 & 0 & -p_y \\ 0 & 0 & 1 & -p_z \\ 0 & 0 & 0 & 1 \end{bmatrix} \begin{bmatrix} n_x & o_x & a_x & p_x \\ n_y & o_y & a_y & p_y \\ n_z & o_z & a_z & p_z \\ 0 & 0 & 0 & 1 \end{bmatrix}
$$

$$
= \begin{bmatrix} n_x & o_x & a_x & 0 \\ n_y & o_y & a_y & 0 \\ n_z & o_z & a_z & 0 \\ 0 & 0 & 0 & 1 \end{bmatrix}
$$

(7.11)

By this translation, we have defined a *hand orientation frame*, $T_{OH}$, (Figure 7.7) whose $x$, $y$, and $z$ axes are exactly parallel to the corresponding axes in the robot base frame. In this frame, we can discuss the orientation of the fingers in a simple way.

Now, we can define, in solving for orientations, *the problem to be solved:*

$T_{OH}$ is given. It is the location of the hand, in homogeneous coordinates, translated from the base. We know $T_{OH}$. We wish to find another way of describing $T_{OH}$, a description in terms of roll, pitch, and yaw. To do this, we will explicitly state the RPY transform and equate it to $T_{OH}$. Then we will solve for the three angles.

### 7.2.1 RPY Representation for Orientation

Roll, pitch, and yaw are defined as rotation about the $x$, $y$, and $z$ axes, respectively, in that order.

$$RPY(\phi_z, \phi_y, \phi_x) = \text{Rot}(z, \phi_z)\,\text{Rot}(y, \phi_y)\,\text{Rot}(x, \phi_x) \qquad (7.12)$$

$$= \begin{bmatrix} c\phi_z & -s\phi_z & 0 & 0 \\ s\phi_z & c\phi_z & 0 & 0 \\ 0 & 0 & 1 & 0 \\ 0 & 0 & 0 & 1 \end{bmatrix} \begin{bmatrix} c\phi_y & 0 & s\phi_y & 0 \\ 0 & 1 & 0 & 0 \\ -s\phi_y & 0 & c\phi_y & 0 \\ 0 & 0 & 0 & 1 \end{bmatrix}$$

$$\times \begin{bmatrix} 1 & 0 & 0 & 0 \\ 0 & c\phi_x & -s\phi_x & 0 \\ 0 & s\phi_x & c\phi_x & 0 \\ 0 & 0 & 0 & 1 \end{bmatrix}$$

$$= \begin{bmatrix} c\phi_z c\phi_y & c\phi_z s\phi_y s\phi_x - s\phi_z c\phi_x & c\phi_z s\phi_y c\phi_x + s\phi_z s\phi_x & 0 \\ s\phi_z c\phi_y & s\phi_z s\phi_y s\phi_x + c\phi_z c\phi_x & s\phi_z s\phi_y c\phi_x - c\phi_z s\phi_x & 0 \\ -s\phi_y & c\phi_y s\phi_x & c\phi_y c\phi_x & 0 \\ 0 & 0 & 0 & 1 \end{bmatrix}$$

If we equate this matrix to $T_{OH}$, we should be able to solve for $\phi_z$, $\phi_y$, and $\phi_x$. That is, if

$$T_{OH} = RPY(\phi_z, \phi_y, \phi_x) \qquad (7.13)$$

we can derive 16 simultaneous equations, which must all be true for the matrix identity to hold. If each equation is true, then the solution appears easy; for example, equating the 3, 1 elements of both matrices, we find

$$-\sin \phi_y = n_z \qquad (7.14)$$

$$\phi_y = \sin^{-1}(-n_z) \qquad (7.15)$$

Then, equating the 3, 2 elements provides

$$\cos \phi_y \sin \phi_x = o_z \qquad (7.16)$$

$$\phi_x = \sin^{-1}\left(\frac{o_z}{\cos \phi_y}\right) \qquad (7.17)$$

This approach seems too easy, and in fact it is, for several reasons (Paul, 1981):

1. Both the inverse sine and inverse cosine functions are not well behaved numerically. The accuracy depends on the angle.
2. The inverse sine and inverse cosine are ambiguous, since $\sin \theta = \sin(\pi - \theta)$.
3. Division by $\cos \phi_y$, as in Eq. 7.17, is inaccurate as $\phi_y \rightarrow \pi$.

For these reasons, we use a slightly different approach. The strategy will be to manipulate the equations in such a way that angles may be found using the inverse tangent. The ambiguity can easily be resolved by using the *ATAN2 function* rather than the conventional ATAN, since ATAN2 takes two arguments and unambiguously returns the inverse tangent in the proper quadrant.

To solve Eq. 7.13 using the ATAN2 function, we premultiply both sides by $\text{Rot}(z, \phi_z)^{-1}$.

$$\text{Rot}(z, \phi_z)^{-1} \cdot T_{OH} = \text{Rot}(y, \phi_y)\,\text{Rot}(x, \phi_x) \tag{7.18}$$

resulting in

$$\begin{bmatrix} f_{11} & f_{12} & f_{13} & 0 \\ f_{21} & f_{22} & f_{23} & 0 \\ f_{31} & f_{32} & f_{33} & 0 \\ 0 & 0 & 0 & 1 \end{bmatrix} = \begin{bmatrix} c\phi_y & s\phi_y s\phi_x & s\phi_y c\phi_x & 0 \\ 0 & c\phi_x & -s\phi_x & 0 \\ -s\phi_y & c\phi_y s\phi_x & c\phi_y c\phi_x & 0 \\ 0 & 0 & 0 & 1 \end{bmatrix} \tag{7.19}$$

where the LHS (left-hand side) is defined as follows:

$$\text{Rot}(z, \phi_z)^{-1} \cdot T_{OH} = \begin{bmatrix} c\phi_z & s\phi_z & 0 & 0 \\ -s\phi_z & c\phi_z & 0 & 0 \\ 0 & 0 & 1 & 0 \\ 0 & 0 & 0 & 1 \end{bmatrix} \begin{bmatrix} n_x & o_x & a_x & 0 \\ n_y & o_y & a_y & 0 \\ n_z & o_z & a_z & 0 \\ 0 & 0 & 0 & 1 \end{bmatrix}$$

$$= \begin{bmatrix} n_x c\phi_z + n_y s\phi_z & o_x c\phi_z + o_y s\phi_z & a_x c\phi_z + a_y s\phi_z & 0 \\ n_y c\phi_z - n_x s\phi_z & o_y c\phi_z - o_x s\phi_z & a_y c\phi_z - a_x s\phi_z & 0 \\ n_z & o_z & a_z & 0 \\ 0 & 0 & 0 & 1 \end{bmatrix}$$

We note that element 2, 1 of the RHS (right-hand side) is zero. Equating $f_{21}$ to 0 yields

$$f_{21} = -n_x \sin \phi_z + n_y \cos \phi_z = 0 \tag{7.20}$$

thus

$$\phi_z = \text{ATAN2}(n_y, n_x) \tag{7.21}$$

Now, equating $f_{11}$ to $\cos \phi_y$ yields

$$\cos \phi_y = n_x \cos \phi_z + n_y \sin \phi_z \tag{7.22}$$

We must avoid the temptation to use the inverse cosine here and instead equate $f_{31}$ to $-\sin \phi_y$:

$$-\sin \phi_y = n_z \tag{7.23}$$

We now divide Eq. 7.23 by Eq. 7.22 to find

$$\phi_y = \text{ATAN2}(-n_z, n_x \cos \phi_z + n_y \sin \phi_z) \tag{7.24}$$

Finally, by a similar equating and division on the 2, 3 and 2, 2 elements, we obtain

$$\phi_x = \text{ATAN2}(a_x \sin \phi_z - a_y \cos \phi_z, o_y \cos \phi_z - o_x \sin \phi_z) \qquad (7.25)$$

**Example 7.5  Determining RPY from a Transform Matrix**

Given that

$$^R T_H = \begin{bmatrix} 1 & 0 & 0 & 0 \\ 0 & 0 & 1 & 5 \\ 0 & -1 & 0 & 3 \\ 0 & 0 & 0 & 1 \end{bmatrix}$$

determine $\phi_x$, $\phi_y$, and $\phi_z$.

**Solution:**    From Eq. 7.21,

$$\phi_z = \text{ATAN2}(n_y, n_x)$$
$$= \text{ATAN2}(0, 1)$$
$$= 0°$$

From Eq. 7.24,

$$\phi_y = \text{ATAN2}(-n_z, n_x \cos \phi_z + n_y \sin \phi_z)$$
$$\text{ATAN2}[0, (1)(1) + (0)(0)] = 0$$

Finally, we evaluate Eq. 7.25,

$$\phi_x = \text{ATAN2}(a_x \sin \phi_z - a_y \cos \phi_z, o_y \cos \phi_z - o_x \sin \phi_z)$$
$$= \text{ATAN2}(-1, 0) = 90°$$

### 7.2.2 Euler Angles

Like the RPY notation, *Euler angles* are defined by rotations about the coordinate axes. In this case, the angles are defined in terms of a rotation about the $z$ axis, then a rotation about the new (rotated) $y$ axis, and finally a rotation about the new $z$ axis.

$$\text{Euler}(\phi, \theta, \psi) = \text{Rot}(z, \phi) \, \text{Rot}(y, \theta) \, \text{Rot}(z, \psi) \qquad (7.26)$$

The Euler angles can be determined from $T_{OH}$ using the ATAN2 function in a manner exactly analogous to that described for RPY.

### 7.2.3 Representations for Orientation

We now have three options for describing position and orientation:

Vector-vector (Euler or RPY)
Vector-matrix
Homogeneous

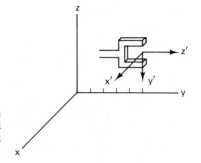

**Figure 7.8** The hand displaced from base coordinates by $[0, 5, 3]$ and oriented in the positive $Y$ direction (rotated $-90°$ about $x$).

## Example 7.6 Equivalent Representations

For the example shown in Figure 7.8 describe the hand configuration in all four forms.

**Solution:**

$$\text{Vector-RPY}: \langle [0, 5, 3], [0, 0, -90] \rangle$$

$$\text{Vector-Euler}: \langle [0, 5, 3], [0, 90, 90] \rangle$$

$$\text{Vector-matrix}: \left\langle [0, 5, 3], \begin{bmatrix} 1 & 0 & 0 \\ 0 & 0 & 1 \\ 0 & -1 & 0 \end{bmatrix} \right\rangle$$

$$\text{Homogeneous}: \begin{bmatrix} 1 & 0 & 0 & 0 \\ 0 & 0 & 1 & 5 \\ 0 & -1 & 0 & 3 \\ 0 & 0 & 0 & 1 \end{bmatrix}$$

These various options for representing orientation provide us with a mechanism for describing the orientation of the hand in a somewhat more intuitive manner than simply writing down the $4 \times 4$ transform. Furthermore, by learning how to solve for the orientation, we have learned a valuable lesson that will be useful in the next section. However, at this point, we need to summarize where we are.

### 7.2.4 Summary: The Forward Kinematic Transform

Using the $A$ matrices, we have developed a technique for going from joint space to Cartesian space (Figure 7.9). That is, we assume

**Figure 7.9** The forward kinematic transform finds the Cartesian configuration corresponding to a given set of joint angles.

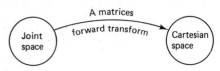

Figure 7.10  The inverse kinematic transform, or "arm solution," assumes that
a position in space is known and finds
joint angles that correspond to that
position.

the joint variables $\theta_1 - \theta_6$ are known (they must be, for they are directly measured), and from that, we compute the configuration of the hand. A joint servo, as discussed in Chapter 5, requires not only $\theta$ but also $\theta_d$. In most practical applications, we must be able to take a given hand configuration (a desired configuration) and find the joint angles that will yield that configuration (Figure 7.10).

This, the inverse problem, is referred to as the *arm solution*. We will demonstrate one example derivation of the arm solution, for an elbow manipulator, in the following section.

## 7.3 THE ARM SOLUTION

In this section, we will demonstrate the inverse kinematic transform, known as the "arm solution." We will initially consider the $\theta$-$r$ manipulator in order to present the general concepts of the inverse transform. Then, we will derive the solution for the articulated manipulator, utilizing the same mathematical approach we used in Section 7.2 to find the orientation.

### 7.3.1  Solution of the $\theta$-$r$ Manipulator

In Section 7.1.1, we developed the forward kinematic transform for the $\theta$-$r$ manipulator, based on an arbitrary assignment of coordinate frames. Once we assigned those frames, we developed $A$ matrices, the product of which related the hand frame to the base frame.

$$^R T_H = A_1 A_2 \tag{7.27}$$

$$= \begin{bmatrix} \cos\theta & -\sin\theta & 0 & 0 \\ \sin\theta & \cos\theta & 0 & 0 \\ 0 & 0 & 1 & 0 \\ 0 & 0 & 0 & 1 \end{bmatrix} \cdot \begin{bmatrix} 1 & 0 & 0 & r \\ 0 & 1 & 0 & 0 \\ 0 & 0 & 1 & 0 \\ 0 & 0 & 0 & 1 \end{bmatrix} \tag{7.28}$$

Now, we are faced with the complementary problem: Given a particular transform $^R T_H$, representing a hand configuration, what choice of $\theta$ and $r$ will result in this hand configuration? That is, we need to solve the following equation for $r$ and $\theta$.

$$\begin{bmatrix} n_x & o_x & a_x & p_x \\ n_y & o_y & a_y & p_y \\ n_z & o_z & a_z & p_z \\ 0 & 0 & 0 & 1 \end{bmatrix} = \begin{bmatrix} c\,\theta & -s\,\theta & 0 & 0 \\ s\,\theta & c\,\theta & 0 & 0 \\ 0 & 0 & 1 & 0 \\ 0 & 0 & 0 & 1 \end{bmatrix} \cdot \begin{bmatrix} 1 & 0 & 0 & r \\ 0 & 1 & 0 & 0 \\ 0 & 0 & 1 & 0 \\ 0 & 0 & 0 & 1 \end{bmatrix}$$

$$(7.29)$$

Initially, we observe that this is unsolvable* in general, for it demands that

$$n_z = o_z = a_x = a_y = p_z = 0 \quad \text{and} \quad a_z = 1$$

This is reasonable, since the $\theta$-$r$ manipulator is restricted to motion in the $z = 0$ plane and no control over the $z$ dimension is possible. If we then allow this restriction, and multiply the two $A$ matrices, we have

$$\begin{bmatrix} n_x & o_x & 0 & p_x \\ n_y & o_y & 0 & p_y \\ 0 & 0 & 1 & 0 \\ 0 & 0 & 0 & 1 \end{bmatrix} = \begin{bmatrix} c\,\theta & -s\,\theta & 0 & rc\theta \\ s\,\theta & c\,\theta & 0 & rs\theta \\ 0 & 0 & 1 & 0 \\ 0 & 0 & 0 & 1 \end{bmatrix}$$

$$(7.30)$$

By equating the 1, 4 and 2, 4 elements, we find

$$r \cos \theta = p_x \qquad (7.31)$$

and

$$r \sin \theta = p_y \qquad (7.32)$$

We take the ratio of these two equations to find

$$\tan \theta = \frac{p_y}{p_x} \qquad (7.33)$$

$$\theta = \text{ATAN2}(p_y, p_x) \qquad (7.34)$$

To find $r$, we premultiply by $A_1^{-1}$,

$$\begin{bmatrix} c\,\theta & s\,\theta & 0 & 0 \\ -s\,\theta & c\,\theta & 0 & 0 \\ 0 & 0 & 1 & 0 \\ 0 & 0 & 0 & 1 \end{bmatrix} \begin{bmatrix} n_x & o_x & 0 & p_x \\ n_y & o_y & 0 & p_y \\ 0 & 0 & 1 & 0 \\ 0 & 0 & 0 & 1 \end{bmatrix} = \begin{bmatrix} 1 & 0 & 0 & r \\ 0 & 1 & 0 & 0 \\ 0 & 0 & 1 & 0 \\ 0 & 0 & 0 & 1 \end{bmatrix}$$

$$(7.35)$$

*This unsolvability is a fundamental concept that is often overlooked by students.

From the 1, 4 term, we find

$$p_x \cos \theta + p_y \sin \theta = r \qquad (7.36)$$

and the problem is solved.

We should be suspicious of this solution, since it is independent of $n_x$, $n_y$, $o_x$, and $o_y$. But, again, we recall that the $\theta$-$r$ manipulator has no control over orientation, only position, and it, therefore, is logical that the solution we derive is independent of the orientation parameters.

**Example 7.7  Solution of the $\theta$-$r$ Manipulator**

Given that

$$
{}^R T_H = \begin{bmatrix}
1 & 0 & 0 & 2 \\
0 & 0 & 1 & 5 \\
0 & -1 & 0 & 3 \\
0 & 0 & 0 & 1
\end{bmatrix}
$$

determine the values of $r$ and $\theta$ that position the hand of the $\theta$-$r$ manipulator so that it best matches this configuration.

**Solution:**  We immediately note that this configuration is unattainable, as the $\theta$-$r$ manipulator is constrained to move only in the $x$-$y$ plane. Therefore, the 3, 4 element, $p_z = 3$, is an impossible condition. Since we cannot move in $z$, we ignore this element and evaluate Eq. 7.34:

$$\theta = \text{ATAN2}(5, 2) = 68°$$

Then, Eq. 7.36 provides

$$r = 2 \cos 68° + 5 \sin 68° = 5.38$$

We note that the orientation specified by the first three rows and columns of this matrix is the same as that of example 7.6, which resulted from a rotation of $-90°$ about $x$. Again, since the $\theta$-$r$ manipulator cannot control orientation, these terms are ignored in the solution.

#### 7.3.2  Solution of an Articulated Manipulator

For a general manipulator with 6 degrees of freedom, we describe the hand configuration in terms of the $A$ matrices by

$$
{}^R T_H = A_1 A_2 A_3 A_4 A_5 A_6 \qquad (7.37)
$$

where the $A$ matrices are given in Figure 7.6 and ${}^R T_H$ is assumed to be known. As we did in the case of solving for orientation, we premultiply by $A_1^{-1}$,

$$
A_1^{-1} \cdot {}^R T_H = A_2 A_3 A_4 A_5 A_6
$$

or

$$\text{LHS} = \text{RHS}$$

where RHS is given by Eq. 7.38 and LHS is given by Eq. 7.39.

$$\text{RHS} = \begin{bmatrix} c_{234}c_5c_6 - s_{234}s_6 & -c_5c_6c_{234} - c_6s_{234} & c_{234}s_5 & a_4c_{234} + a_3c_{23} + a_2c_2 \\ s_{234}c_5c_6 + c_{234}s_6 & c_{234}c_6 - s_{234}c_5c_6 & s_{234}s_5 & a_4s_{234} + a_3s_{23} + a_2s_2 \\ -s_5c_6 & s_5s_6 & c_5 & 0 \\ 0 & 0 & 0 & 1 \end{bmatrix}$$

$$(7.38)$$

where the elements of the LHS are derived by multiplying $^R T_H$ by $A_1^{-1}$. (Recall the definition of the elements of $^R T_H$ given in Eq. 7.5.)

$$\text{LHS} = \begin{bmatrix} c_1 & s_1 & 0 & 0 \\ 0 & 0 & 1 & 0 \\ s_1 & -c_1 & 0 & 0 \\ 0 & 0 & 0 & 1 \end{bmatrix} \cdot \begin{bmatrix} n_x & o_x & a_x & p_x \\ n_y & o_y & a_y & p_y \\ n_z & o_z & a_z & p_z \\ 0 & 0 & 0 & 1 \end{bmatrix}$$

$$\therefore \text{LHS} = \begin{bmatrix} n_xc_1 + n_ys_1 & o_xc_1 + o_ys_1 & a_xc_1 + a_ys_1 & p_xc_1 + p_ys_1 \\ n_z & o_z & a_z & p_z \\ n_yc_1 - n_xs_1 & o_yc_1 - o_xs_1 & a_yc_1 - a_xs_1 & p_yc_1 - p_xs_1 \\ 0 & 0 & 0 & 1 \end{bmatrix}$$

$$(7.39)$$

We equate the individual elements of the matrices and look for conditions whereby we can solve for an angle using the ATAN2 function. Initially, we observe the 3, 4 elements involve only $\theta_1$,

$$-\sin\theta_1\, p_x + \cos\theta_1\, p_y = 0$$

$$\theta_1 = \tan^{-1}\left(\frac{p_y}{p_x}\right) \qquad (7.40)$$

Now that we have $\theta_1$, we look for an equation relating some other angle to $\theta_1$ and discover the 3, 3 element, resulting in

$$a_y\cos\theta_1 - a_x\sin\theta_1 = \cos\theta_5 \qquad (7.41)$$

Having now found a solution for one of the angles and the cosine of another, further inspection fails to yield any more useful equations. We generate a new set of equations by premultiplying by $A_2^{-1}$.

$$A_2^{-1} \cdot A_1^{-1} \cdot {}^R T_H = A_3 A_4 A_5 A_6 \qquad (7.42)$$

But this turns out to provide no useful information (this occurs because the joints are parallel). In fact, we find nothing useful until we reach a nonparallel link, $A_4$, and

$$A_4^{-1} \cdot A_3^{-1} \cdot A_2^{-1} \cdot A_1^{-1} \cdot {}^R T_H = A_5 \cdot A_6 \qquad (7.43)$$

or

$$\begin{bmatrix} c_{234}(n_x c_1 + n_y s_1) + n_z s_{234} & c_{234}(o_x c_1 + o_y s_1) + o_z s_{234} & c_{234}(a_x c_1 + a_y s_1) + a_z s_{234} & c_{234}(p_x c_1 + p_y s_1) + o_z s_{234} - a_2 c_{34} - a_3 c_4 - a_4 \\ n_y c_1 - n_x s_1 & o_y c_1 - o_x s_1 & a_y c_1 - a_x s_1 & 0 \\ n_z c_{234} - s_{234}(n_x c_1 + n_y s_1) & o_z c_{234} - s_{234}(o_x c_1 + o_y s_1) & a_z c_{234} - s_{234}(a_x c_1 + a_y s_1) & c_{234} p_z - s_{234}(p_x c_1 + p_y s_1) + a_2 s_{34} + a_3 s_4 \\ 0 & 0 & 0 & 1 \end{bmatrix}$$

$$= \begin{bmatrix} c_5 c_6 & -c_5 c_6 & s_5 & 0 \\ s_5 c_6 & -s_5 c_6 & -c_5 & 0 \\ s_6 & c_6 & 0 & 0 \\ 0 & 0 & 0 & 1 \end{bmatrix} \qquad (7.44)$$

We can find (by careful inspection) that the 3, 3 element provides an equation involving only $s_{234}$ and functions of $\theta_1$:

$$a_z c_{234} - s_{234}(a_x c_1 + a_y s_1) = 0 \qquad (7.45)$$

so

$$(\theta_2 + \theta_3 + \theta_4) = \text{ATAN2}(a_z, a_x c_1 + a_y s_1) \qquad (7.46)$$

We now know $\theta_1$ and $\theta_{234}$, and we look for a relationship including these. From Eq. 7.38, we find, from the 1, 4 elements

$$p_x c_1 + p_y s_1 = a_4 c_{234} + a_3 c_{23} + a_2 c_2 \qquad (7.47)$$

and from the 2, 4 elements, we find

$$p_z = a_4 s_{234} + a_3 s_{23} + a_2 s_2 \qquad (7.48)$$

We redefine those combinations of known parameters to make the algebra simpler:

$$p_1 = p_x c_1 + p_y s_1 - a_4 c_{234} \qquad (7.49)$$

$$p_2 = p_z - a_4 s_{234} \qquad (7.50)$$

We then solve Eqs. 7.47 and 7.48 simultaneously to find

$$p_1 = a_3 c_{23} + a_2 c_2 \qquad (7.51)$$

$$p_2 = a_3 s_{23} + a_2 s_2 \qquad (7.52)$$

$$c_3 = \frac{p_1^2 + p_2^2 - a_3^2 - a_2^2}{2 a_2 a_3} \qquad (7.53)$$

and

$$\theta_3 = \text{ATAN2} \{\sqrt{(1 - c_3^2)}, c_3\} \tag{7.54}$$

Now, simultaneous solutions of Eqs. 7.51 and 7.52 will yield

$$S_2 = \frac{(a_3 c_3 + a_2) p_2 - a_3 s_3 p_1}{(a_3 c_3 + a_2)^2 + a_3^2 s_3^2} \tag{7.55}$$

and

$$C_2 = \frac{(a_3 c_3 + a_2) p_1 + a_3 s_3 p_2}{(a_3 c_3 + a_2)^2 + a_3^2 s_3^2} \tag{7.56}$$

and

$$\theta_2 = \text{ATAN2} \{[(a_3 c_3 + a_2) p_2 - a_3 s_3 p_1], [(a_3 c_3 + a_2) p_1 + a_3 s_3 p_2]\} \tag{7.57}$$

Now, we know $\theta_2$, $\theta_3$, and $\theta_{234}$. Therefore, we can find $\theta_4$ by

$$\theta_4 = \theta_{234} - (\theta_2 + \theta_3) \tag{7.58}$$

In Eq. 7.44, we observe that the 1, 3 element contains only one unknown, $s_5$.

$$s_5 = c_{234}(a_x c_1 + a_y s_1) + a_z s_{234} \tag{7.59}$$

and the 2, 3 element likewise contains only $c_5$.

$$-c_5 = a_y c_1 - a_x s_1 \tag{7.60}$$

$$\theta_5 = \text{ATAN2} \{[c_{234}(a_x c_1 + a_y s_1) + a_z s_{234}], (a_x s_1 - a_y c_1)\} \tag{7.61}$$

At this point, neither Eq. 7.38 nor Eq. 7.44 helps us to find $\theta_6$, so we premultiply by $A_5^{-1}$ as before to produce

$$\begin{bmatrix} c_5[c_{234}(n_x c_1 + n_y s_1) + n_z s_{234}] - s_5(n_x s_1 - n_y c_1) & c_5[c_{234}(o_x c_1 + o_y s_1) + o_z s_{234}] - s_5(o_x s_1 - o_y c_1) & 0 & 0 \\ n_z c_{234} - s_{234}(n_x c_1 + n_y s_1) & o_z c_{234} - s_{234}(o_x c_1 + o_y c_1) & 0 & 0 \\ 0 & 0 & 1 & 0 \\ 0 & 0 & 0 & 1 \end{bmatrix}$$

$$= \begin{bmatrix} c_6 & -s_6 & 0 & 0 \\ s_6 & c_6 & 0 & 0 \\ 0 & 0 & 1 & 0 \\ 0 & 0 & 0 & 1 \end{bmatrix} \tag{7.62}$$

and we solve for $\theta_6$ as

$$\theta_6 = \text{ATAN2} \{ [s_5 (o_x s_1 - o_y c_1) - c_5 [c_{234}(o_x c_1 + o_y s_1) + o_z s_{234}]], $$
$$[o_z c_{234} - s_{234}(o_x c_1 + o_y s_1)]\} \qquad (7.63)$$

## 7.4 PROBLEMS IN COMPUTING KINEMATICS

Two phenomena may occur that impact our calculations of kinematic transforms. These are known as degeneracies and singularities. A *degeneracy* occurs when more than one solution to the inverse transform exists. A *singularity* occurs when a denominator in one of the transform calculations goes to zero. We will examine these two phenomena separately, using simplified manipulators to present the basic concepts. Degeneracies will be discussed in this section. Singularities will be covered in the next chapter, as we discuss transformations of velocities.

### 7.4.1 Degeneracies

Whenever there is more than one set of joint variables that result in the same hand configuration, the arm is said to be *degenerate*. Several conditions may lead to a degeneracy. We will consider only one here, the state of joint 3 of an elbow manipulator.

Recall from Eq. 7.54 that the square root is taken in calculating $\theta_3$. The positive sign of the solution was arbitrarily chosen. If, however, the negative sign had been chosen, it would have still provided a legitimate solution, but with the elbow down rather than up.

Such degeneracies often occur, calling for ad hoc decisions on which solution to choose. In general, no major difficulties occur in resolving such ambiguities.

A more significant problem occurs in arms with more than six degrees of freedom. In such arms, it is possible that there can exist an infinite set of possible combinations of joint variables which result in the same hand configurations. Such arms are referred to as *infinitely degenerate*. Humans, as they have a large number of degrees of freedom, possess this property.

With infinitely degenerate arms, other techniques must be employed to solve the kinematics, such as minimum energy optimization, techniques that are beyond the scope of this course. Of course, one can always fix one of the actuators into an unmoving state and thereby reduce the number of degrees of freedom to six.

## 7.5 CONCLUSION

The techniques shown in this chapter, particularly those involving the inverse transform, may seem to be extremely ad hoc. In some sense, this is true, for there does not exist at this time an algorithm for obtaining closed-form solutions to the inverse transforms for arbitrary arm geometries. Iterative, numerical solutions can always be obtained, but at such computational expense as to make them infeasible, particularly for real-time use. The strategy presented here works in most cases and results in closed form solutions. The strategy/philosophy may be summarized as follows.

1. Represent the (given) hand transform $^R T_H$ as the product of the $A$ matrices.
2. Search for relationships involving the sine and cosine of a single joint variable. These relations should provide the ability to calculate angles using the inverse tangent.
3. Iteratively premultiply both sides of the matrix equations by inverses of $A$ matrices to form new sets of equations on which to apply step 2.

## 7.6 SYNOPSIS

### Vocabulary

You should know the definition and application of the following terms:

$A$ matrix
arm solution
ATAN2
base coordinates
degeneracy
Euler angle
forward kinematic transform
hand orientation frame
inverse kinematic transform
joint
link
prismatic joint
rotary joint
RPY
twist angle

**Notation**

| Symbol | Meaning |
| --- | --- |
| $A_i$ | The $4 \times 4$ matrix relating the coordinate system for joint $i$ to that of joint $i-1$ |
| $^R T_H$ | The $4 \times 4$ matrix relating the coordinate system of the hand to that of the robot's base |
| $\alpha$ | Twist angle |
| $\theta_i$ | The joint variable, usually a rotation, associated with the $i$th joint |
| $\theta$ | The first joint of the $\theta$-$r$ manipulator ($\theta_1$) |
| $r$ | The second joint of the $\theta$-$r$ manipulator ($\theta_2$) |
| $s_i$ | Abbreviation for sin ($\theta_i$) |
| $c_i$ | Abbreviation for cos ($\theta_i$) |
| $s_{ijk}$ | Abbreviation for sin ($\theta_i + \theta_j + \theta_k$) |
| $c_{ijk}$ | Abbreviation for cos ($\theta_i + \theta_j + \theta_k$) |
| n | Normal vector, the first column of $^R T_H$ |
| o | Orientation vector, the second column of $^R T_H$ |
| a | Approach vector, the third column of $^R T_H$ |
| p | Position vector, fourth column of $^R T_H$ |
| $n_x$ | The first element of n and the 11 element of $^R T_H$ |
| $n_y$ | The second element of n and the 21 element of $^R T_H$ |
| $n_z$ | The third element of n and the 31 element of $^R T_H$ |
| $o_x, o_y, o_z$ | Defined analogous to $n_x$ |
| $a_x, a_y, a_z$ | Defined analogous to $n_x$ |
| $p_x, p_y, p_z$ | Defined analogous to $n_x$ |
| $T_{OH}$ | Hand orientation frame |
| $\phi_z$ | Rotation about the $z$ axis (roll) |
| $\phi_y$ | Rotation about the $y$ axis (pitch) |
| $\phi_x$ | Rotation about the $x$ axis (yaw) |
| RPY | A $4 \times 4$ matrix resulting from combinations of roll, pitch, and yaw |

## 7.7 REFERENCES

SANDOR, B. I. *Engineering Mechanics: Design*. Englewood Cliffs, N.J.: Prentice-Hall, 1983, p. 499.

PAUL, R. P. *Robot Manipulators*. Cambridge, Mass.: MIT Press, 1981.

## 7.8 PROBLEMS

1. Find the inverses of the matrices given in Figure 7.6.
2. Show that Eq. 7.62 can be solved to result in Eq. 7.63.
3. Show that Eqs. 7.51 and 7.52 are the solutions to the simultaneous solution of Eqs. 7.47 and 7.48.
4. You will be presented with a robot that utilizes a geometry very similar to the articulated manipulator. As a group, measure this arm, determine the $A$ matrices and the transform, $^R T_H$.
5. For the arm in problem 4, find the arm solution.

## APPENDIX: COORDINATE FRAME DEFINITIONS—PRISMATIC JOINTS

Figure 7A.1 shows a spherical manipulator that incorporates both rotary and prismatic joints. Joint 3 of this arm is prismatic. To determine the appropriate $A$ matrices, we must assign coordinate frames at both ends of the link. Figure 7A.2 details the links of this arm. When assigning a frame for joint 2, we locate the origin at the end of the link as before, but choose the positive direction of $z$ to be aligned with the direction of motion of the prismatic joint. The origin of joint 3 is chosen to be

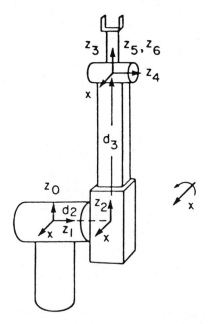

**Figure 7A.1**   (Paul) A manipulator utilizing a spherical geometry but in which the centers of rotation for joints 1 and 2 are offset by the distance $d_2$.

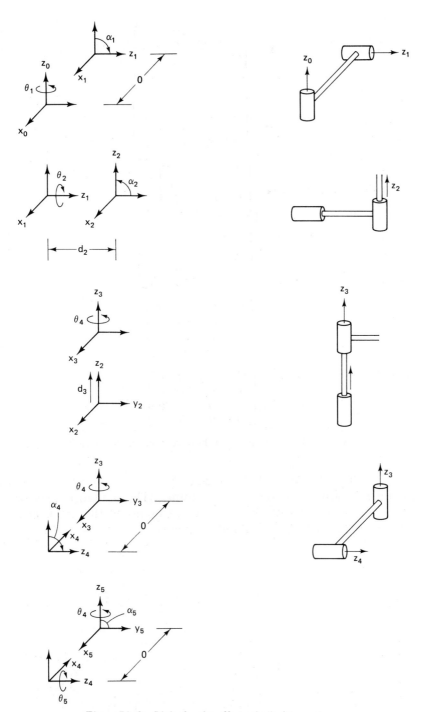

**Figure 7A.2**  Links for the offset spherical geometry.

$$A_1 = \begin{bmatrix} C_1 & 0 & -S_1 & 0 \\ S_1 & 0 & C_1 & 0 \\ 0 & -1 & 0 & 0 \\ 0 & 0 & 0 & 1 \end{bmatrix}$$

$$A_2 = \begin{bmatrix} C_2 & 0 & S_2 & 0 \\ S_2 & 0 & -C_2 & 0 \\ 0 & 1 & 0 & d_2 \\ 0 & 0 & 0 & 1 \end{bmatrix}$$

$$A_3 = \begin{bmatrix} 1 & 0 & 0 & 0 \\ 0 & 1 & 0 & 0 \\ 0 & 0 & 1 & d_3 \\ 0 & 0 & 0 & 1 \end{bmatrix}$$

$$A_4 = \begin{bmatrix} C_4 & 0 & -S_4 & 0 \\ S_4 & 0 & C_4 & 0 \\ 0 & -1 & 0 & 0 \\ 0 & 0 & 0 & 1 \end{bmatrix}$$

$$A_5 = \begin{bmatrix} C_5 & 0 & S_5 & 0 \\ S_5 & 0 & -C_5 & 0 \\ 0 & 1 & 0 & 0 \\ 0 & 0 & 0 & 1 \end{bmatrix}$$

$$A_6 = \begin{bmatrix} C_6 & -S_6 & 0 & 0 \\ S_6 & C_6 & 0 & 0 \\ 0 & 0 & 1 & 0 \\ 0 & 0 & 0 & 1 \end{bmatrix}$$

**Figure 7A.3** $A$ matrices for the offset spherical geometry.

located at the other end of the prismatic link, with the $z$ axis aligned with the rotational axis of the next joint. The $A$ matrices shown in Figure 7A.3 result for this arm. It should be noted that $A_3$ is now reduced to a simple translation along the $z$ axis.

# 8

---

# DIFFERENTIAL
# MOTIONS AND
# THE JACOBIAN

In this chapter we will be concerned with the kinematic transformations as they relate to velocities. In Chapter 9, we will learn that in many of the more sophisticated robot control techniques, we must be able to determine, and possibly control, the velocity of the hand. We include the time derivatives of both position and orientation in our velocity determination. We will assume that we have some knowledge of joint velocities, either by direct measurement or by calculation of rotational displacement per unit time. As we did in Chapter 7 with displacement, we will determine the transformations that relate joint velocities to Cartesian velocities, and vice versa. As we also did in Chapter 7, we will discuss only rotary joints. We remember, however, that prismatic joints also exist and have a motion which is described by a translation.

In any digital system, measurements are made at discrete intervals of time which are each separated by a finite, nonzero sampling time, and estimates cannot be updated more frequently. Hence, one can never know the true velocity but can only make measurements of displacements over the sampling time, and estimate the velocity from these measurements. If our system is designed properly, this estimate will be very close to the true instantaneous velocity.

If the sampling time is a constant, and we will for the moment assume that is the case, then the problem of determining velocity can be reduced to determining the small *differential* motion which occurs during the sampling interval.

## 8.1 DIFFERENTIAL MOTION

We will begin our analysis of differential motions by considering the single-axis rotation matrices:

$$\text{Rot}(x, \theta) = \begin{bmatrix} 1 & 0 & 0 & 0 \\ 0 & \cos\theta & -\sin\theta & 0 \\ 0 & \sin\theta & \cos\theta & 0 \\ 0 & 0 & 0 & 1 \end{bmatrix} \tag{8.1}$$

$$\text{Rot}(y, \theta) = \begin{bmatrix} \cos\theta & 0 & \sin\theta & 0 \\ 0 & 1 & 0 & 0 \\ -\sin\theta & 0 & \cos\theta & 0 \\ 0 & 0 & 0 & 1 \end{bmatrix} \tag{8.2}$$

$$\text{Rot}(z, \theta) = \begin{bmatrix} \cos\theta & -\sin\theta & 0 & 0 \\ \sin\theta & \cos\theta & 0 & 0 \\ 0 & 0 & 1 & 0 \\ 0 & 0 & 0 & 1 \end{bmatrix} \tag{8.3}$$

As we are considering differential motions, we define $\delta_x$ to be a small rotation about the $x$ axis. For small angles, $\sin\theta = \theta$, $\cos\theta = 1$. We substitute these values into Eqs. 8.1, 8.2, and 8.3 to find

$$\text{Rot}(x, \delta_x) = \begin{bmatrix} 1 & 0 & 0 & 0 \\ 0 & 1 & -\delta_x & 0 \\ 0 & \delta_x & 1 & 0 \\ 0 & 0 & 0 & 1 \end{bmatrix} \tag{8.4}$$

$$\text{Rot}(y, \delta_y) = \begin{bmatrix} 1 & 0 & \delta_y & 0 \\ 0 & 1 & 0 & 0 \\ -\delta_y & 0 & 1 & 0 \\ 0 & 0 & 0 & 1 \end{bmatrix} \tag{8.5}$$

$$\text{Rot}(z, \delta_z) = \begin{bmatrix} 1 & -\delta_z & 0 & 0 \\ \delta_z & 1 & 0 & 0 \\ 0 & 0 & 1 & 0 \\ 0 & 0 & 0 & 1 \end{bmatrix} \tag{8.6}$$

A general differential rotation can be found by multiplying Eqs. 8.4–8.6 together

$\text{Rot}(x, \delta_x) \, \text{Rot}(y, \delta_y) \, \text{Rot}(z, \delta_z)$

$$= \begin{bmatrix} 1 & 0 & \delta_y & 0 \\ \delta_x \delta_y & 1 & -\delta_x & 0 \\ -\delta_y & \delta_x & 1 & 0 \\ 0 & 0 & 0 & 1 \end{bmatrix} \cdot \begin{bmatrix} 1 & -\delta_z & 0 & 0 \\ \delta_z & 1 & 0 & 0 \\ 0 & 0 & 1 & 0 \\ 0 & 0 & 0 & 1 \end{bmatrix}$$

(8.7)

$$= \begin{bmatrix} 1 & -\delta_z & \delta_y & 0 \\ \delta_x \delta_y + \delta_z & 1 - \delta_x \delta_y \delta_z & -\delta_x & 0 \\ (-\delta_y + \delta_x \delta_z) & (\delta_y \delta_z + \delta_x) & 1 & 0 \\ 0 & 0 & 0 & 1 \end{bmatrix}$$

which, again, we simplify by neglecting second and third order terms.

$$\begin{bmatrix} 1 & -\delta_z & \delta_y & 0 \\ \delta_z & 1 & -\delta_x & 0 \\ -\delta_y & \delta_x & 1 & 0 \\ 0 & 0 & 0 & 1 \end{bmatrix}$$

(8.8)

**Theorem 8.1**

A differential rotation $\delta_\theta$ about an arbitrary vector $\mathbf{K} = [k_x, k_y, k_z]^\mathsf{T}$ is equivalent to three differential rotations, $\delta_x, \delta_y, \delta_z$, about the $x, y,$ and $z$ axes, where

$$k_x \delta_\theta = \delta_x$$

$$k_y \delta_\theta = \delta_y$$

(8.9)

$$k_z \delta_\theta = \delta_z$$

The proof may be found in Paul (1981b).

It is interesting to note the following theorem.

**Theorem 8.2**

Differential rotations are independent of the order in which the rotations are performed. This comes about because of the assumption in which we ignored the second- and third-order terms. We prove it for only one case:

$$\text{Rot}(x, \delta_x) \, \text{Rot}(y, \delta_y) = \text{Rot}(y, \delta_y) \, \text{Rot}(x, \delta_x)$$

(8.10)

154

That is,

$$
\begin{bmatrix} 1 & 0 & 0 & 0 \\ 0 & 1 & -\delta_x & 0 \\ 0 & \delta_x & 1 & 0 \\ 0 & 0 & 0 & 1 \end{bmatrix} \cdot \begin{bmatrix} 1 & 0 & \delta_y & 0 \\ 0 & 1 & 0 & 0 \\ -\delta_y & 0 & 1 & 0 \\ 0 & 0 & 0 & 1 \end{bmatrix}
$$

$$
\stackrel{?}{=} \begin{bmatrix} 1 & 0 & \delta_y & 0 \\ 0 & 1 & 0 & 0 \\ -\delta_y & 0 & 1 & 0 \\ 0 & 0 & 0 & 1 \end{bmatrix} \cdot \begin{bmatrix} 1 & 0 & 0 & 0 \\ 0 & 1 & -\delta_x & 0 \\ 0 & \delta_x & 1 & 0 \\ 0 & 0 & 0 & 1 \end{bmatrix}
$$

By multiplying matrices, this becomes

$$
\begin{bmatrix} 1 & 0 & \delta_y & 0 \\ \delta_x\delta_y & 1 & -\delta_x & 0 \\ -\delta_y & \delta_x & 1 & 0 \\ 0 & 0 & 0 & 1 \end{bmatrix} \stackrel{?}{=} \begin{bmatrix} 1 & \delta_x\delta_y & \delta_y & 0 \\ 0 & 1 & -\delta_x & 0 \\ -\delta_y & \delta_x & 1 & 0 \\ 0 & 0 & 0 & 1 \end{bmatrix}
$$

Again, we assume that second-order terms may be neglected and set $\delta_x\delta_y$ to zero. The two matrices are then equal.

We can define a *differential motion* as a combined translation and rotation. That is,

$$
T + dT = [\text{Trans}(dx, dy, dz)\, \text{Rot}(\mathbf{K}, d\theta)]\, T \tag{8.11}
$$

where $\mathbf{K}$ is the arbitrary axis of rotation discussed in theorem 8.1 and $T$ is a homogeneous transform matrix.

Then

$$
dT = \text{Trans}(dx, dy, dz)\, \text{Rot}(\mathbf{K}, d\theta)\, T - T \tag{8.12}
$$

Factoring out $T$, we find

$$
dT = [\text{Trans}(dx, dy, dz)\, \text{Rot}(\mathbf{K}, d\theta) - I]\, T \tag{8.13}
$$

Thus, if we augment the generalized rotation matrix in Eq. 8.8 by adding the translation vector and subtracting the identity matrix, we define a *differential operator*

$$
\Delta = \begin{bmatrix} 0 & -\delta_z & \delta_y & dx \\ \delta_z & 0 & -\delta_x & dy \\ -\delta_y & \delta_x & 0 & dz \\ 0 & 0 & 0 & 0 \end{bmatrix} = \text{Trans}(dx, dy, dz)\, \text{Rot}(\mathbf{K}, d\theta) - I
$$

$$
\tag{8.14}
$$

where $[dx, dy, dz]^T$ represents a differential translation and $\delta_x$, $\delta_y$, and $\delta_z$ are differential rotations.

Given a hand configuration described by a homogeneous transformation $T$, we can determine the results of applying a differential
change to that configuration simply by premultiplying by $\Delta$.

$$dT = \Delta T \qquad (8.15)$$

**Example 8.1: Effect of Differential Motion**

Suppose that the hand configuration is described by

$$T = \begin{bmatrix} 0 & 0 & 1 & 5 \\ 1 & 0 & 0 & 2 \\ 0 & 1 & 0 & 0 \\ 0 & 0 & 0 & 1 \end{bmatrix}$$

What is the effect of a differential rotation of 0.1 rad (radians) about the $x$ axis,
followed by a differential translation of $[1, 0, 0.5]^T$?

**Solution:**    We first determine that

$$\begin{aligned} \delta_x &= 0.1 & dx &= 1 \\ \delta_y &= 0.0 & dy &= 0 \\ \delta_z &= 0.0 & dz &= 0.5 \end{aligned}$$

and substitute into Eq. 8.14

$$\Delta = \begin{bmatrix} 0 & 0 & 0 & 1 \\ 0 & 0 & -0.1 & 0 \\ 0 & 0.1 & 0 & 0.5 \\ 0 & 0 & 0 & 0 \end{bmatrix}$$

Thus, the result of this differential motion,

$$dT = \Delta T = \begin{bmatrix} 0 & 0 & 0 & 1 \\ 0 & 0 & -0.1 & 0 \\ 0 & 0.1 & 0 & 0.5 \\ 0 & 0 & 0 & 0 \end{bmatrix} \cdot \begin{bmatrix} 0 & 0 & 1 & 5 \\ 1 & 0 & 0 & 2 \\ 0 & 1 & 0 & 0 \\ 0 & 0 & 0 & 1 \end{bmatrix} = \begin{bmatrix} 0 & 0 & 0 & 1 \\ 0 & -0.1 & 0 & 0 \\ 0.1 & 0 & 0 & 0.7 \\ 0 & 0 & 0 & 0 \end{bmatrix}$$

We must emphasize that the principles of differential motion are
based upon the assumption that the differential motions are small. As
we begin to describe the *instantaneous velocity*, the notion of small
differential motions will be very appropriate. However, we must be
careful when we make the small-motion assumption to calculate the
effects of motions relative to the current position. Making use of that
assumption has allowed us to convert the complex equation (6.15) to
the much simpler form of Eq. 8.11. In the foregoing example, $\delta_\theta$ was
0.1 rad.

We can see the effects of the small-angle assumption in the following table.

| | Rule | True Value | Error |
|---|---|---|---|
| $\theta = 0.1$ | $\sin(\theta) = \theta$ | 0.0998334 | 0.0001666 |
| $\theta = 0.1$ | $\cos(\theta) = 1$ | 0.995004 | 0.004996 |

Thus, even for differential motions as large as 0.1 rad (5.7°), the results are quite accurate. If we know where we are, we can determine where we will be after a small movement by applying a simple operator, involving no transcendental functions.

We cannot, of course, use this technique iteratively since errors accumulate. Instead, we will use it to learn how to transform velocities.

## 8.2 THE JACOBIAN

We could define the differential displacement of the hand as a vector with six elements:

$$\mathbf{D} = \begin{bmatrix} dx \\ dy \\ dz \\ \delta_x \\ \delta_y \\ \delta_z \end{bmatrix} \tag{8.16}$$

where, as before, $\delta_x$ represents a rotation about the $x$ axis.

Similarly, we could define a six-element vector consisting of the differential displacements of the six joint variables.

$$\mathbf{D}_\theta = \begin{bmatrix} d\theta_1 \\ d\theta_2 \\ d\theta_3 \\ d\theta_4 \\ d\theta_5 \\ d\theta_6 \end{bmatrix} \tag{8.17}$$

where, as before, the symbol $\theta$ may represent either a rotation (for rotary joints) or an extension (for prismatic joints).

Use of these two vectors defines the 6 × 6 *Jacobian matrix, J:*

$$\mathbf{D} = J\mathbf{D}_\theta \tag{8.18}$$

The Jacobian thus relates velocities in joint space to velocities in Cartesian space. And if the inverse exists,

$$\mathbf{D}_\theta = J^{-1}\mathbf{D} \tag{8.19}$$

Calculation of the Jacobian and its inverse for a six-degrees-of-freedom arm is quite complex. Before we move into that calculation, we will derive these matrices for the two-degrees-of-freedom $\theta$-$r$ manipulator.

### 8.2.1 The Jacobian of the $\theta$-$r$ Manipulator

For the $\theta$-$r$ manipulator, the Cartesian displacement vector $\mathbf{D}$ can be written

$$\mathbf{D} = \begin{bmatrix} dx \\ dy \end{bmatrix} \tag{8.20}$$

As we discussed earlier, we are often interested in velocities, and we observe that a differential displacement which occurs over a *constant* differential time $\Delta t_{samp}$ can be considered as estimating an instantaneous velocity.

Thus,

$$\mathbf{D} = [\dot{x}, \dot{y}]^{\mathsf{T}} \tag{8.21}$$

In terms of the joint variables, $r$ and $\theta$, the velocity vector is written

$$\mathbf{D}_\theta = [\dot{r}, \dot{\theta}]^{\mathsf{T}} \tag{8.22}$$

The Jacobian is then defined by

$$\mathbf{D} = J\mathbf{D}_\theta$$
$$\begin{bmatrix} \dot{x} \\ \dot{y} \end{bmatrix} = \begin{bmatrix} J_{11} & J_{12} \\ J_{21} & J_{22} \end{bmatrix} \begin{bmatrix} \dot{r} \\ \dot{\theta} \end{bmatrix} \tag{8.23}$$

We know that

$$x = r \cos \theta$$
$$y = r \sin \theta$$

Taking derivatives with respect to time

$$\frac{dx}{dt} = \left(\frac{\partial r}{\partial t}\right) \cos \theta + r \left(\frac{\partial}{\partial t}\right) \cos \theta \tag{8.24}$$
$$= \dot{r} \cos \theta - \dot{\theta} \, r \sin \theta$$

$$\frac{dy}{dt} = \frac{\partial r}{\partial t} \sin \theta + r \frac{\partial}{\partial t} \sin \theta \tag{8.25}$$
$$= \dot{r} \sin \theta + \dot{\theta} \, r \cos \theta$$

Thus, we can write

$$\begin{bmatrix} \dot{x} \\ \dot{y} \end{bmatrix} = \begin{bmatrix} \cos\theta & -r\sin\theta \\ \sin\theta & r\cos\theta \end{bmatrix} \cdot \begin{bmatrix} \dot{r} \\ \dot{\theta} \end{bmatrix} \tag{8.26}$$

and

$$J(r,\theta) = \begin{bmatrix} \cos\theta & -r\sin\theta \\ \sin\theta & r\cos\theta \end{bmatrix} \tag{8.27}$$

We find the inverse Jacobian in a similar way, by writing

$$r^2 = x^2 + y^2$$

and taking the derivative with respect to time

$$2r\frac{dr}{dt} = 2x\frac{dx}{dt} + 2y\frac{dy}{dt} \tag{8.28}$$

Therefore, $\dot{r} = \dfrac{x}{r}\,\dot{x} + \dfrac{y}{r}\,\dot{y}$

We find the second row using

$$\tan\theta = \frac{y}{x}$$

and, taking the derivative with respect to time.

$$\dot{\theta}\sec^2\theta = \frac{-y}{x^2}\,\dot{x} + \frac{1}{x}\,\dot{y} \tag{8.29}$$

$$\dot{\theta} = \frac{-y}{x^2(r^2/x^2)}\,\dot{x} + \frac{1}{x(r^2/x^2)}\,\dot{y} = \frac{-y}{r^2}\,\dot{x} + \frac{x}{r^2}\,\dot{y}$$

And we have

$$\begin{bmatrix} \dot{r} \\ \dot{\theta} \end{bmatrix} = \begin{bmatrix} x/r & y/r \\ -y/r^2 & x/r^2 \end{bmatrix} \cdot \begin{bmatrix} \dot{x} \\ \dot{y} \end{bmatrix} \tag{8.30}$$

and

$$J_{(r,\theta)}^{-1} = \begin{bmatrix} x/r & y/r \\ -y/r^2 & x/r^2 \end{bmatrix} \tag{8.31}$$

Two observations are in order at this point regarding the derivatives of the Jacobian for the $\theta$-$r$ manipulator. First, the reader should note that the inverse Jacobian contains $r$ in the denominator of several terms. From this, we can conclude that something "unpleasant" happens at $r = 0$. A careful analysis of this case reveals that for constant Cartesian velocity ($\dot{x} = c$), the angular velocity, $\dot{\theta}$, must approach infinity if the path passes through $r = 0$. Whenever a joint velocity is required to become infinite to maintain constant Cartesian velocity, we have a *singularity*. Manipulator control is lost at singularities. Arm configurations that lead to singularities must therefore be explicitly avoided. Fortunately, these conditions show up as zeros in the denominator of terms in the inverse Jacobian, or as conditions that make the determinant of the Jacobian zero, and are, therefore, easy to identify.

The second observation regards the method of the derivation. In finding both $J$ and $J^{-1}$, we have taken derivatives with respect to time and then manipulated the algebra to find a convenient expression. This manipulation was reasonably straightforward for this simple, two-joint manipulator. In the more general case, with six joints, the algebraic manipulation becomes increasingly difficult, and a more structured approach is desired. This approach will be addressed in the next section.

**Example 8.2: Effects of Singularities on the Jacobian**

Using the inverse Jacobian presented in Eq. 8.30, determine $\dot{\theta}$ as a function of $x$ as the hand moves along the line $y = 1$ at a uniform speed of $\dot{x} = 1$.

**Solution:**    Since $y = 1$, $\dot{y} = 0$, and the equation for $\dot{\theta}$ becomes

$$\dot{\theta} = \frac{-\dot{x}}{r^2} = \frac{-1}{x^2 + 1}$$

We note that $\dot{\theta}$ is a function that has a peak at $x = 0$.

### 8.2.2 Transforming Differential Motions Between Frames

A good technique for determining the Jacobian of a six-degrees-of-freedom manipulator is to make use of the $A$ matrices which define the geometry. We essentially reflect the velocities through the $A$ matrices. To do this, we must learn the way in which velocities or differential motions relate from one frame to another.

More specifically, in Section 8.1 we defined a differential motion (or velocity) operator $\Delta$ which, when applied to a transform matrix $T$, would give us the resulting differential motion. That is, $dT = \Delta T$, where $\Delta$ is the $4 \times 4$ differential motion operator. But that operator is defined in the same frame as $T$. Another important problem arises when we observe the motion from another frame. That is, given a differential motion defined in one frame, what is that same motion when viewed in another frame?

To approach this problem, we define the differential motion transform relative to an arbitrary frame $T$ as $^T\Delta$ and we can characterize the differential motion in base* coordinates as

---

*Henceforth, we will assume a universe frame coincident with the base of the manipulator. Reference to this frame will be identified by the *absence* of a superscript on the left of a transform. Thus, $\Delta = {}^U\Delta$.

$$dT = T^T\Delta \tag{8.32}$$

Equating Eqs. 8.15 and 8.32, we find

$$\Delta T = T^T\Delta \tag{8.33}$$

or

$$^T\Delta = T^{-1}\Delta T$$

We define the following variables to simplify the notation:

$$\delta = [\delta_x, \delta_y, \delta_z]^T$$
$$n = [n_x, n_y, n_z]^T$$
$$o = [o_x, o_y, o_z]^T$$
$$a = [a_x, a_y, a_z]^T$$
$$p = [p_x, p_y, p_z]^T$$

As before, let $T$ be defined by

$$T = \begin{bmatrix} n_x & o_x & a_x & p_x \\ n_y & o_y & a_y & p_y \\ n_z & o_z & a_z & p_z \\ 0 & 0 & 0 & 1 \end{bmatrix}$$

We find $\Delta T$ by

$$= \begin{bmatrix} 0 & -\delta_z & \delta_y & d_x \\ \delta_z & 0 & -\delta_x & d_y \\ -\delta_y & \delta_x & 0 & d_z \\ 0 & 0 & 0 & 0 \end{bmatrix} \begin{bmatrix} n_x & o_x & a_x & p_x \\ n_y & o_y & a_y & p_y \\ n_z & o_z & a_z & p_z \\ 0 & 0 & 0 & 1 \end{bmatrix} \tag{8.34}$$

$$\Delta T = \begin{bmatrix} -\delta_z n_y + \delta_y n_z & -\delta_z o_y + \delta_y o_z & -\delta_z a_y + \delta_y a_z & -\delta_z p_y + \delta_y p_z + dx \\ \delta_z n_x - \delta_x n_z & \delta_z o_x - \delta_x o_z & \delta_z a_x - \delta_x a_z & \delta_z p_x - \delta_x p_z + dy \\ \delta_x n_y - \delta_y n_x & \delta_x o_y - \delta_y o_x & \delta_x a_y - \delta_y a_x & \delta_x p_y - \delta_y p_x + dz \\ 0 & 0 & 0 & 0 \end{bmatrix}$$

$$\tag{8.35}$$

At this point we multiply by $T^{-1}$ to find $^T\Delta = T^{-1}\Delta T$. We have three choices, multiply out all the terms, search for a more compact notation, or multiply out only those terms which we will need. We choose the third option. To determine which terms we need, we go back to the original derivation of $\Delta$ (Eq. 8.14) and rewrite that result, making explicit the fact that it is defined *in frame T*.

$$^T\Delta = \begin{bmatrix} 0 & -^T\delta_z & ^T\delta_y & ^Tdx \\ ^T\delta_z & 0 & -^T\delta_x & ^Tdy \\ -^T\delta_y & ^T\delta_x & 0 & ^Tdz \\ 0 & 0 & 0 & 0 \end{bmatrix} \tag{8.36}$$

We then equate this form for $^T\Delta$ to Eq. 8.35 and solve for the differential motion terms.

$$T^{-1}\,\Delta T = \begin{bmatrix} n_x & n_y & n_z & -\mathbf{n}\cdot\mathbf{p} \\ o_x & o_y & o_z & -\mathbf{o}\cdot\mathbf{p} \\ a_x & a_y & a_z & -\mathbf{a}\cdot\mathbf{p} \\ 0 & 0 & 0 & 1 \end{bmatrix}$$

$$\times \begin{bmatrix} -\delta_z n_y + \delta_y n_z & -\delta_z o_y + \delta_y o_z & -\delta_z a_y + \delta_y a_z & -\delta_z p_y + \delta_y p_z + dx \\ \delta_z n_x - \delta_x n_z & \delta_z o_x - \delta_x o_z & \delta_z a_x - \delta_x a_z & \delta_z p_x - \delta_x p_z + dy \\ \delta_x n_y - \delta_y n_x & \delta_x o_y - \delta_y o_x & \delta_x a_y - \delta_y a_x & \delta_x p_y - \delta_y p_x + dz \\ 0 & 0 & 0 & 0 \end{bmatrix}$$

$$= \begin{bmatrix} 0 & -^T\delta_z & ^T\delta_y & ^Tdx \\ ^T\delta_z & 0 & -^T\delta_x & ^Tdy \\ -^T\delta_y & ^T\delta_x & 0 & ^Tdz \\ 0 & 0 & 0 & 0 \end{bmatrix} \tag{8.37}$$

Solve for the (2, 1) term:

$$^T\delta_z = o_x(-\delta_z n_y + \delta_y n_z) + o_y(\delta_z n_x - \delta_x n_z) + o_z(\delta_x n_y - \delta_y n_x) \tag{8.38}$$

Solve for the (1, 3) term:

$$^T\delta_y = n_x(-\delta_z a_y + \delta_y a_z) + n_y(\delta_z a_x - \delta_x a_z) + n_z(\delta_x a_y - \delta_y a_x) \tag{8.39}$$

Solve for the (3, 2) term:

$$^T\delta_x = a_x(-\delta_z o_y + \delta_y o_z) + a_y(\delta_z o_x - \delta_x o_z) + a_z(\delta_x o_y - \delta_y o_x) \tag{8.40}$$

Similarly,

$$^Tdx = n_x(-\delta_z p_y + \delta_y p_z + dx) + n_y(\delta_z p_x - \delta_x p_z + dy)$$
$$+ n_z(\delta_x p_y - \delta_y p_x + dz) \tag{8.41}$$

$$^Tdy = o_x(-\delta_z p_y + \delta_y p_z + dx) + o_y(\delta_z p_x - \delta_x p_z + dy)$$
$$+ o_z(\delta_x p_y - \delta_y p_x + dz) \tag{8.42}$$

$$^Tdz = a_x(-\delta_z p_y + \delta_y p_z + dx) + a_y(\delta_z p_x - \delta_x p_z + dy)$$
$$+ a_z(\delta_x p_y - \delta_y p_x + dz) \tag{8.43}$$

Equation (8.38) can be simplified as follows: Rearranging terms

$$^T\delta_z = \delta_x(o_z n_y - o_y n_z) + \delta_y(o_x n_z - o_z n_x) + \delta_z(o_y n_x - o_x n_y) \tag{8.44}$$

which can be written as the dot product of two vectors:

$$^T\delta_z = \begin{bmatrix} \delta_x \\ \delta_y \\ \delta_z \end{bmatrix}^T \cdot \begin{bmatrix} o_z n_y - o_y n_z \\ o_x n_z - o_z n_x \\ o_y n_x - o_x n_y \end{bmatrix} \tag{8.45}$$

We recognize the second vector to be the cross product of the vectors $[n_x\ n_y\ n_z]^T$ and $[o_x\ o_y\ o_z]^T$

$$^T\delta_z = \delta \cdot (\mathbf{n} \times \mathbf{o}) \qquad (8.46)$$

but since

$$\mathbf{n} \times \mathbf{o} = \mathbf{a} \qquad (8.47)$$

we determine that

$$^T\delta_z = \delta \cdot \mathbf{a} \qquad (8.48)$$

Similarly, we can simplify Eqs. 8.39–8.43 to yield

$$^T\delta_y = \delta \cdot \mathbf{o} \qquad (8.49)$$

$$^T\delta_x = \delta \cdot \mathbf{n} \qquad (8.50)$$

$$^Tdx = \mathbf{n} \cdot [(\delta \times \mathbf{p}) + \mathbf{d}] \qquad (8.51)$$

$$^Tdy = \mathbf{o} \cdot [(\delta \times \mathbf{p}) + \mathbf{d}] \qquad (8.52)$$

$$^Tdz = \mathbf{a} \cdot [(\delta \times \mathbf{p}) + \mathbf{d}] \qquad (8.53)$$

Thus we have derived a technique for computing the differential motions as viewed from different frames. We will make use of this result in the next section.

### 8.2.3 The Manipulator Jacobian

From Eq. 8.18, we have

$$\mathbf{D} = J\mathbf{D}_\theta$$

We now can make explicit the relationship between differential motions and velocities by defining

$$\dot{\mathbf{X}} = \left(\frac{1}{\Delta t_{samp}}\right)\mathbf{D}$$

where $\dot{\mathbf{X}}$ is the velocity, expressed in Cartesian coordinates and $\Delta t_{samp}$ is the scalar-valued sampling interval. We define the elements of $\dot{\mathbf{X}}$ by $\dot{\mathbf{X}} = [\dot{x}\ \dot{y}\ \dot{z}\ \dot{\phi}_x\ \dot{\phi}_y\ \dot{\phi}_z]^T$, where the $\phi$'s are rotational velocities. Furthermore, we define $\theta = [\dot{\theta}_1\ \dot{\theta}_2\ \dot{\theta}_3\ \dot{\theta}_4\ \dot{\theta}_5\ \dot{\theta}_6]^T$. Then,

$$
\begin{bmatrix} \dot{x} \\ \dot{y} \\ \dot{z} \\ \dot{\phi}_x \\ \dot{\phi}_y \\ \dot{\phi}_z \end{bmatrix}
=
\begin{bmatrix}
J_{11} & J_{12} & J_{13} & J_{14} & J_{15} & J_{16} \\
J_{21} & J_{22} & J_{23} & J_{24} & J_{25} & J_{26} \\
J_{31} & J_{32} & J_{33} & J_{34} & J_{35} & J_{36} \\
J_{41} & J_{42} & J_{43} & J_{44} & J_{45} & J_{46} \\
J_{51} & J_{52} & J_{53} & J_{54} & J_{55} & J_{56} \\
J_{61} & J_{62} & J_{63} & J_{64} & J_{65} & J_{66}
\end{bmatrix}
\begin{bmatrix} \dot{\theta}_1 \\ \dot{\theta}_2 \\ \dot{\theta}_3 \\ \dot{\theta}_4 \\ \dot{\theta}_5 \\ \dot{\theta}_6 \end{bmatrix}
\qquad (8.54)
$$

where we have deliberately changed the notation from differential motions to velocities to emphasize the equivalence of these concepts for constant sampling time.

The objective of this section is to determine the coefficients of the $J$ matrix from the known kinematics (the $A$ matrices) of the manipulator.

Immediately, we have a notational problem. As we have written it, $\dot{\mathbf{X}}$ represents velocities relative to the base frame, including rotational velocities about the axes of the base frame. We could equally well write

$$^H\dot{\mathbf{X}} = {}^HJ\dot{\boldsymbol{\theta}} \qquad (8.55)$$

Now, $^HJ$ represents a Jacobian matrix that relates joint velocities to instantaneous Cartesian velocities relative to the hand frame. $^HJ$ turns out to be somewhat easier to compute in general. To see why, we will begin by attempting to evaluate $J$.

First, we multiply the first row of the $J$ matrix by $\boldsymbol{\theta}$ to derive

$$\dot{x} = J_{11}\dot{\theta}_1 + J_{12}\dot{\theta}_2 + J_{13}\dot{\theta}_3 + J_{14}\dot{\theta}_4 + J_{15}\dot{\theta}_5 + J_{16}\dot{\theta}_6 \qquad (8.56)$$

From the form of this expression, we see that the elements of the Jacobian are in fact the partial derivatives. That is,

$$J_{11} = \frac{\partial x}{\partial \theta_1} \qquad (8.57a)$$

$$J_{12} = \frac{\partial x}{\partial \theta_2}, \text{etc} \qquad (8.57b)$$

That being the case, we should be able to find the elements of the first row of $J$ by taking the derivative with respect to time of an expression relating $x$ to the joint variables. The kinematic transform is exactly such an expression. The term $p_x$ of the hand transform is the position of the hand in space along the $x$ axis. The derivatives of $p_x$ with respect to time will yield the $x$ component of velocity.

From Eq. (7.9) we have

$$p_x = c_1[a_4 c_{234} + a_3 c_{23} + a_2 c_2] \qquad (7.9)$$

$$\frac{dp_x}{dt} = c_1[-a_4 s_{234}(\dot{\theta}_2 + \dot{\theta}_3 + \dot{\theta}_4) - a_3 s_{23}(\dot{\theta}_2 + \dot{\theta}_3) - a_2 s_2 \dot{\theta}_2]$$

$$- [a_4 c_{234} + a_3 c_{23} + a_2 c_2] s_1 \dot{\theta}_1 \qquad (8.58)$$

$$\frac{dp_x}{dt} = -(a_4 s_1 c_{234} + a_3 s_1 c_{23} + a_2 s_1 c_2)\dot{\theta}_1$$

$$- (a_4 c_1 s_{234} + a_3 s_{23} + a_2 s_2)\dot{\theta}_2$$

$$- (a_4 c_1 s_{234} + a_3 s_{23})\dot{\theta}_3 \qquad (8.59)$$

$$- (a_4 c_1 s_{234})\dot{\theta}_4$$

Therefore,

$$J_{11} = \frac{\partial x}{\partial \theta_1} = -(a_4 s_1 c_{234} + a_3 s_1 c_{23} + a_2 s_1 c_2) \qquad (8.60a)$$

$$J_{12} = \frac{\partial x}{\partial \theta_2} = -(a_4 c_1 s_{234} + a_3 s_{23} + a_2 s_2) \qquad (8.60b)$$

$$J_{13} = \frac{\partial x}{\partial \theta_3} = -(a_4 c_1 s_{234} + a_3 s_{23}) \qquad (8.60c)$$

$$J_{14} = \frac{\partial x}{\partial \theta_4} = -(a_4 c_1 s_{234}) \qquad (8.60d)$$

$$J_{15} = \frac{\partial x}{\partial \theta_5} = 0 \qquad (8.60e)$$

$$J_{16} = \frac{\partial x}{\partial \theta_6} = 0 \qquad (8.60f)$$

A similar approach can be used to find the elements of the second and third row of the Jacobian by differentiating $p_y$ and $p_z$, respectively.

We have more difficulty, however, with the fourth through sixth rows. This problem occurs because the hand transform $^U T_H$ does not contain an explicit term for the rotational variables. For example, $\phi_z$ represents rotation about the $z$ axis (this is as defined in Section 7.2.1). $\dot{\phi}_z$ is then the rotational velocity about the $z$ axis and is equivalent to $\delta_z$. In Section 7.2.1, we developed a method of finding an explicit form for $\phi_x$, $\phi_y$, and $\phi_z$ from the elements of $T_H$. [In particular, $\phi_z = \tan^{-1}(n_y/n_x)$.] We could differentiate this expression with respect to time, resulting in an expression involving the elements of $T_H$ and their derivatives. Equation (7.9) could then be differentiated and used to find the sixth row of $J$. However, the mathematics are tedious, and the form of the resultant is cumbersome. Instead, we will use a different approach, in which we will find $^H J$.

### An Algorithm for Finding $^H J$

We present only the algorithm here. The derivation may be found in Paul (1981b).

First, we remind the reader of the convention for establishing joint coordinate frames ($A$ matrices) defined in Section 7.1. Those conventions are

1.  For rotary joints, the joint variable $\theta_i$ corresponds to a rotation about the $z$ axis of the previous link frame, $z_{i-1}$.

**2.** For prismatic joints, the joint variable $r_i$ corresponds to a translation along the $z$ axis of the previous frame, $a_{i-1}$.

With these conventions, we establish a hand transform $T_H$ which is the product of six $A$ matrices and we name the elements of $T_H$:

$$T_H = \begin{bmatrix} n_x & o_x & a_x & p_x \\ n_y & o_y & a_y & p_y \\ n_z & o_z & a_z & p_z \\ 0 & 0 & 0 & 1 \end{bmatrix}$$

We also define five intermediate frames by

$$^1T_H = A_1^{-1} T_H$$
$$^2T_H = A_2^{-1} A_1^{-1} T_H$$
$$^3T_H = A_3^{-1} A_2^{-1} A_1^{-1} T_H$$
$$^4T_H = A_4^{-1} A_3^{-1} A_2^{-1} A_1^{-1} T_H$$
$$^5T_H = A_5^{-1} A_4^{-1} A_3^{-1} A_2^{-1} A_1^{-1} T_H = A_6$$

When referencing an element of one of these matrices, we use the left superscript to designate the matrix, for example, $n_x$ refers to the $(1, 1)$ term of $T_H$, but $^2n_x$ refers to the $(1, 1)$ term of $^2T_H$.

The elements of the hand frame Jacobian $^HJ$ may be found as follows: For rotary joints, where $\theta_i$ is the joint variable,

$$\frac{^H\partial x}{\partial \theta_{i+1}} = {}^i n_y {}^i p_x - {}^i n_x {}^i p_y \qquad (8.61a)$$

$$\frac{^H\partial y}{\partial \theta_{i+1}} = {}^i o_y {}^i p_x - {}^i o_x {}^i p_y \qquad (8.61b)$$

$$\frac{^H\partial z}{\partial \theta_{i+1}} = {}^i a_y {}^i p_x - {}^i a_x {}^i p_y \qquad (8.61c)$$

$$\frac{^H\partial \phi_z}{\partial \theta_{i+1}} = {}^i n_z \qquad (8.61d)$$

$$\frac{^H\partial \phi_y}{\partial \theta_{i+1}} = {}^i o_z \qquad (8.61e)$$

$$\frac{^H\partial \phi_x}{\partial \theta_{i+1}} = {}^i a_z \qquad (8.61f)$$

For prismatic joints, where $r_i$ is the joint variable

$$\frac{^H\partial x}{\partial r_{i+1}} = {}^i n_z \qquad (8.62a)$$

$$\frac{^H\partial_y}{\partial r_{i+1}} = {}^i o_z \tag{8.62b}$$

$$\frac{^H\partial_z}{\partial r_{i+1}} = {}^i a_z \tag{8.62c}$$

$$\frac{^H\partial \phi_z}{\partial r_i} = \frac{^H\partial \phi_y}{\partial r_i} = \frac{^H\partial \phi_x}{\partial r_i} = 0 \tag{8.62d}$$

### The Jacobian of the Articulated Manipulator

The articulated manipulator has no prismatic joints, so we will use Eq. 8.61 exclusively. We will derive only two of the terms symbolically here, to demonstrate the technique.

First, we list the $T$ matrices: $T_H$ is given in Eq. 7.9.

$${}^1T_H = A_2 A_3 A_4 A_5 A_6 = A_1{}^{-1} T_H$$

$$= \begin{bmatrix} c_{234}c_5c_6 - s_{234}s_6 & -c_{234}c_5c_6 - s_{234}c_6 & c_{234}s_5 & c_{234}a_4 + c_{23}a_3 + c_2a_2 \\ s_{234}c_5c_6 + c_{234}s_6 & -s_{234}c_5s_6 + c_{234}c_6 & s_{234}s_5 & s_{234}a_4 + s_{23}a_3 + s_2a_2 \\ -s_5c_6 & s_5s_6 & c_5 & 0 \\ 0 & 0 & 0 & 1 \end{bmatrix}$$

$$\tag{8.63a}$$

$${}^2T_H = A_3 A_4 A_5 A_6 = A_2^{-1}\, {}^1T_H$$

$$= \begin{bmatrix} c_{34}c_5c_6 - s_{34}s_6 & -c_{34}c_5s_6 - s_{34}c_6 & c_{34}s_5 & c_{34}a_4 + c_3a_3 \\ s_{34}c_5c_6 + c_{34}s_6 & -s_{34}c_5s_6 + c_{34}c_6 & s_{34}s_5 & s_{34}a_4 + s_3a_3 \\ -s_5c_6 & s_5s_6 & c_5 & 0 \\ 0 & 0 & 0 & 1 \end{bmatrix}$$

$$\tag{8.63b}$$

$${}^3T_H = A_4 A_5 A_6 = A_3^{-1}\, {}^2T_H$$

$$= \begin{bmatrix} c_4c_5c_6 - s_4s_6 & -c_4c_5s_6 - s_4c_6 & c_4s_5 & c_4a_4 \\ s_4c_5c_6 + c_4s_6 & -s_4c_5s_6 + c_4c_6 & s_4s_5 & s_4a_4 \\ -s_5c_6 & s_5s_6 & c_5 & 0 \\ 0 & 0 & 0 & 1 \end{bmatrix} \tag{8.63c}$$

$$^4T_H = A_4^{-1} \; ^3T_H$$

$$= \begin{bmatrix} c_5 c_6 & -c_5 s_6 & s_5 & 0 \\ s_5 c_6 & -s_5 s_6 & -c_5 & 0 \\ s_6 & c_6 & 0 & 0 \\ 0 & 0 & 0 & 1 \end{bmatrix} \tag{8.63d}$$

$$^5T_H = A_6$$

$$= \begin{bmatrix} c_6 & -s_6 & 0 & 0 \\ s_6 & c_6 & 0 & 0 \\ 0 & 0 & 1 & 0 \\ 0 & 0 & 0 & 1 \end{bmatrix} \tag{8.63e}$$

We apply Eq. 8.61a to $T_H$ to find $J_{11}$:

$$^HJ_{11} = \frac{^H\partial_x}{\partial\theta_1} = n_y p_x - n_x p_y$$

$$= \{s_1 c_{234} c_5 c_6 - s_{234} s_1 s_6 + c_1 s_5 c_6\}$$

$$\cdot \{c_1 c_{234} a_4 + c_1 c_{23} a_3 + c_1 c_2 a_2\} \tag{8.64}$$

$$- \{c_1 c_5 c_6 c_{234} - c_1 s_6 s_{234} - s_1 s_5 c_6\}$$

$$\cdot \{s_1 c_{234} a_4 + s_1 c_{23} a_3 + s_1 c_2 a_2\}$$

We apply Eq. 8.61c to $^3T_H$ to find

$$^HJ_{34} = \frac{^H\partial_z}{\partial\theta_4} = {}^3a_y\,{}^3p_x - {}^3a_x\,{}^3p_y$$

$$= s_4 s_5 c_4 a_4 - c_4 s_5 s_4 a_4 \tag{8.65}$$

All the terms of the Jacobian may be found in an analogous manner.

### Properties of the Jacobian

Several observations are in order with regard to the Jacobian at this point. First, it is not a constant matrix, but depends on the arm configuration. This fact makes real-time computation a major challenge. Second, whereas $J$ relates velocities to the stationary base frame, $^HJ$ relates velocities to a *moving frame*. The student may find this fact confusing. However, the motion of the reference frame is not considered in the expression. The Cartesian velocities such as $dx/dt$ should be considered as measured relative to the instantaneous *position* of the hand frame.

**Example 8.2:** Calculation of $^H J$

Suppose that the articulated manipulator is in the following configuration:

| Variable | Value |
|----------|-------|
| $\theta_1$ | $0°$ |
| $\theta_2$ | $90°$ |
| $\theta_3$ | $0°$ |
| $\theta_4$ | $90°$ |
| $\theta_5$ | $0°$ |
| $\theta_6$ | $90°$ |

Application of Eq. 7.9 to this set of joint variables results in the following $T_H$

$$T_H = \begin{bmatrix} 0 & 1 & 0 & -a_4 \\ 0 & 0 & -1 & 0 \\ -1 & 0 & 0 & a_2 + a_3 \\ 0 & 0 & 0 & 1 \end{bmatrix} \tag{8.66a}$$

In a realistic implementation, we would solve in closed form for the elements of $^H J$, as was done in Eqs. 8.64 and 8.65. The computer would then simply substitute values for the joint variables. In this example, however, we will explicitly evaluate all the $T$ matrices, to demonstrate numerically the order of operations.

$$^1 T_H = \begin{bmatrix} 0 & 1 & 0 & -a_4 \\ -0 & 0 & 0 & a_2 + a_3 \\ 0 & 0 & 1 & 0 \\ 0 & 0 & 0 & 1 \end{bmatrix} \tag{8.66b}$$

$$^2 T_H = \begin{bmatrix} -1 & 0 & 0 & a_3 \\ 0 & -1 & 0 & a_4 \\ 0 & 0 & 1 & 0 \\ 0 & 0 & 0 & 1 \end{bmatrix} \tag{8.66c}$$

$$^3 T_H = \begin{bmatrix} -1 & 0 & 0 & 0 \\ 0 & -1 & 0 & a_4 \\ 0 & 0 & 1 & 0 \\ 0 & 0 & 0 & 1 \end{bmatrix} \tag{8.66d}$$

$$^4 T_H = \begin{bmatrix} 0 & -1 & 0 & 0 \\ 0 & 0 & -1 & 0 \\ 1 & 0 & 0 & 0 \\ 0 & 0 & 0 & 1 \end{bmatrix} \tag{8.66e}$$

$$^5 T_H = \begin{bmatrix} 0 & -1 & 0 & 0 \\ 1 & 0 & 0 & 0 \\ 0 & 0 & 1 & 0 \\ 0 & 0 & 0 & 1 \end{bmatrix} \tag{8.66f}$$

$$H_J = \begin{bmatrix} 0 & a_4 & a_4 & a_4 & 0 & 0 \\ 0 & -(a_2 + a_3) & -a_3 & 0 & 0 & 0 \\ a_4 & 0 & 0 & 0 & 0 & 0 \\ -1 & 0 & 0 & 0 & 1 & 0 \\ 0 & 0 & 0 & 0 & 0 & 0 \\ 0 & 1 & 1 & 1 & 0 & 1 \end{bmatrix} \tag{8.67}$$

Suppose that we have a differential change in joint variables.

$$\mathbf{D}_\theta = \begin{bmatrix} 0.1 \\ 0.1 \\ 0.2 \\ 0.1 \\ 0.1 \\ 0.3 \end{bmatrix} \tag{8.68}$$

Then, the hand frame will undergo a differential displacement (let $a_4 = 20$ cm, $a_3 = 1$ m, $a_2 = 1$ m):

$$\begin{bmatrix} 0.08 \\ -0.40 \\ 0.02 \\ 0.00 \\ 0.00 \\ 0.70 \end{bmatrix} = \begin{bmatrix} 0 & 0.2 & 0.2 & 0.2 & 0 & 0 \\ 0 & -2 & -1 & 0 & 0 & 0 \\ 0.2 & 0 & 0 & 0 & 0 & 0 \\ -1 & 0 & 0 & 0 & 1 & 0 \\ 0 & 0 & 0 & 0 & 0 & 0 \\ 0 & 1 & 1 & 1 & 0 & 1 \end{bmatrix} \cdot \begin{bmatrix} 0.1 \\ 0.1 \\ 0.2 \\ 0.1 \\ 0.1 \\ 0.3 \end{bmatrix} \tag{8.69}$$

This differential motion occurs relative to the hand frame. Recall from Eq. 8.36 that any differential motion can be used to define a differential transform matrix in the same frame.

$$^H\Delta = \begin{bmatrix} 0 & -0.70 & 0 & 0.08 \\ 0.70 & 0 & 0 & -0.40 \\ 0 & 0 & 0 & 0.02 \\ 0 & 0 & 0 & 0 \end{bmatrix} \tag{8.70}$$

We can find the motion in base coordinates as the change in $T_H$ (Eq. 8.32).

$$dT_H = T_H{}^H\Delta$$

$$= \begin{bmatrix} 0 & 1 & 0 & -0.20 \\ 0 & 0 & -1 & 0 \\ -1 & 0 & 0 & 2 \\ 0 & 0 & 0 & 1 \end{bmatrix} \cdot \begin{bmatrix} 0 & -0.70 & 0 & 0.08 \\ 0.70 & 0 & 0 & -0.4 \\ 0 & 0 & 0 & 0.02 \\ 0 & 0 & 0 & 0 \end{bmatrix}$$

$$= \begin{bmatrix} 0.70 & 0 & 0 & -0.4 \\ 0 & 0 & 0 & -0.02 \\ 0 & 0.70 & 0 & -0.08 \\ 0 & 0. & 0 & 0 \end{bmatrix} \tag{8.71}$$

## 8.3 THE INVERSE JACOBIAN

The 6 × 6 Jacobian matrix relates the velocity of the hand to the angular velocities of the joints by

$$\dot{\mathbf{X}} = J\dot{\theta} \tag{8.72}$$

where the vector $\dot{\mathbf{X}}$ represents the six-element velocity vector $[\dot{x}, \dot{y}, \dot{z}, \dot{\phi}_x, \dot{\phi}_y, \dot{\phi}_z]^T$. Where $\dot{\phi}_x$, $\dot{\phi}_y$, and $\dot{\phi}_z$ represent rotational velocities about the $x$, $y$, and $z$ axes, respectively. (Note that Eq. 8.72 is identical to Eq. 8.18; we are simply using a notation of velocity rather than differential displacement.) The inverse Jacobian is used whenever desired hand velocity is specified, and we must compute the desired joint velocities, according to

$$\dot{\theta} = J^{-1}\dot{\mathbf{X}} \tag{8.73}$$

In Section 8.2.1, the inverse Jacobian matrix for the $\theta$-$r$ manipulator was computed. In that case, the Jacobian is a 2 × 2 matrix; however, in the general case, the Jacobian is 6 × 6, and the inverse is nontrivial to compute. In this section, we will discuss the options available to the system designer, when he or she must develop a technique for finding the inverse Jacobian.

### 8.3.1 Numerical Inversion

In Eqs. 8.61 and 8.62, an algorithm for finding the elements of the Jacobian matrix is given. Those elements are complex trigonometric functions of the joint angles, resulting in a matrix of staggering complexity, if one were to write it all out in one place. However, at any single instant in time, the arm is in a single configuration, and all those elements have unique numerical values. Thus, at any given instant, the Jacobian is an array of 36 numbers.

The literature in numerical analysis provides techniques for inverting such matrices. Such techniques must not only be fast, but they must also maintain numerical precision through the calculation. Naïve approaches to inversion can result in serious loss of precision, requiring floating point arithmetic (with its loss of speed) to avoid producing inaccurate results. Inversion strategies such as *Gaussian elimination* and others (see Stewart, 1973, for a good discussion) attempt to maintain precision through the calculation.

Now that we have given consideration to precision, the one remaining difficulty with numerical inversion is speed. In general, if a problem can be solved and simplified analytically, calculation of the answer directly from the analytic solution will be faster than a numerical solution. Let us look more at the inversion of the Jacobian from this point of view.

First, we observe that the problem to be solved is

$$\dot{X} = J\dot{\theta} \qquad (8.74)$$

where the elements of $\dot{X}$ are Cartesian velocities and are known and the elements of $\dot{\theta}$ are angular velocities and are unknown. Symbolically, we write the solution as $\dot{\theta} = J^{-1}\,\dot{X}$, but in fact, we solve Eq. 8.74 directly as a system of linear equations. We never explicitly compute $J^{-1}$. The technique known as Gaussian elimination is often applied to such problems. It consists of reducing the matrix $J$ to upper triangular form and then back substituting to find the solution. The first step, triangulation, requires approximately (Stewart, 1973)

$$\frac{6^3}{3} - \frac{(12)6^2}{2} + 6 \cdot 6 \cdot 5 = 72 \text{ multiplications}$$

The back substitution step takes approximately $(\frac{1}{2})n^2$ or 18 multiplications, for a total of 90 multiplications.

### 8.3.2 Symbolic Inversion

If we leave the Jacobian matrix in its original, symbolic form, we can attempt to find the inverse of that form, using algebra. The complexity of the Jacobian makes this approach either very difficult or impossible, although some solutions have been found in this way (Renaud, 1980).

### 8.3.3 Differentiation of the Inverse Kinematic Solution

In Section 8.2.1 this approach was followed to find the inverse of the Jacobian for the $\theta$-$r$ manipulator. It results in a closed-form solution which is efficient to compute.

The inverse kinematic solution provides a relationship between the joint angles and the hand transform, such that the required joint angles may be computed from a given hand configuration. For example, Eq. 7.40 provides

$$\tan \theta_1 = \frac{p_y}{p_x} \tag{7.40}$$

We restructure this equation to make the mathematics simpler:

$$\frac{\sin \theta_1}{\cos \theta_1} = \frac{p_y}{p_x} \tag{8.75}$$

$$p_x \sin \theta_1 = p_y \cos \theta_1 \tag{8.76}$$

Then, we differentiate with respect to time.

$$p_x \cos \theta_1 \dot{\theta}_1 + \sin \theta_1 \dot{p}_x = -p_y \sin \theta_1 \dot{\theta}_1 + \cos \theta_1 \dot{p}_y \tag{8.77}$$

or

$$(p_x \cos \theta_1 + p_y \sin \theta_1) \dot{\theta}_1 = \cos \theta_1 \dot{p}_y - \sin \theta_1 \dot{p}_x \tag{8.78}$$

$$\dot{\theta}_1 = \frac{\cos \theta_1 \dot{p}_y - \sin \theta_1 \dot{p}_x}{p_x \cos \theta_1 + p_y \sin \theta_1}. \tag{8.79}$$

Similarly, the derivative of other joints may be taken to determine relationships between hand and joint velocities. To see how to apply this derivation, let us consider an example:

**Example 8.4: Computation of Joint Velocities**

Compute the joint velocities corresponding to a translational velocity of

$$\begin{bmatrix} \dot{x} \\ \dot{y} \\ \dot{z} \end{bmatrix} = \begin{bmatrix} 0.2 \\ 0.4 \\ 0.6 \end{bmatrix} \tag{8.80}$$

and rotational velocity of

$$\begin{bmatrix} \dot{\phi}_x \\ \dot{\phi}_y \\ \dot{\phi}_z \end{bmatrix} = \begin{bmatrix} 0 \\ 0.1 \\ 0 \end{bmatrix}$$

given that the articulated manipulator is in the following state:

$$^R T_H = \begin{bmatrix} 0 & 1 & 0 & 5 \\ 1 & 0 & 0 & 2 \\ 0 & 0 & -1 & 0 \\ 0 & 0 & 0 & 1 \end{bmatrix} \tag{8.81}$$

Using Eq. 8.14, we derive $^R \Delta$:

$$^R \Delta = \begin{bmatrix} 0 & 0 & 0.1 & 0.2 \\ 0 & 0 & 0 & 0.4 \\ -0.1 & 0 & 0 & 0.6 \\ 0 & 0 & 0 & 0 \end{bmatrix} \tag{8.82}$$

We find $dT$ by premultiplying by $^R T_H$:

$$dT = \begin{bmatrix} 0 & 1 & 0 & 5 \\ 1 & 0 & 0 & 2 \\ 0 & 0 & -1 & 0 \\ 0 & 0 & 0 & 1 \end{bmatrix} \cdot \begin{bmatrix} 0 & 0 & 0.1 & 0.2 \\ 0 & 0 & 0 & 0.4 \\ -0.1 & 0 & 0 & 0.6 \\ 0 & 0 & 0 & 0 \end{bmatrix} \tag{8.83}$$

$$= \begin{bmatrix} 0 & 0 & 0 & 0.4 \\ 0 & 0 & 0.1 & 0.2 \\ 0.1 & 0 & 0 & -0.6 \\ 0 & 0 & 0 & 0 \end{bmatrix}$$

From this, we have

$$\dot{p}_x = 0.4$$
$$\dot{p}_y = 0.2$$

From the kinematic transform of Eq. 7.40, we have

$$\theta = \tan^{-1} \frac{p_y}{p_x} = \tan^{-1} \frac{2}{5} = 0.381 \text{ rad}$$

$$\dot{\theta} = \frac{\cos{(0.381)}{(0.2)} - \sin{(0.381)}{(0.4)}}{5 \cos{(03.81)} + 2 \sin{(0.381)}}$$

$$= 0.007 \; \text{rad}/\Delta t_{samp}$$

## 8.4 CONCLUSION

In this chapter, we initially developed a strategy for representing differential motions. That strategy was based on the fact that, since the differential motions were small, the approximations

$$\sin \theta = \theta \quad \text{and} \quad \cos \theta = 1$$

could be assumed true. Using these assumptions, we derived a mechanism for computing the differential motions in one frame, as they are seen from another frame.

One further observation was needed to make the transition from differential motion to velocity: differential motion measured over a small sampling time approximates a velocity. With that observation, we changed notation from differential motion to velocity.

The Jacobian matrix was then defined and derived, relating the angular velocities to the hand velocity.

Three different techniques for computing the inverse Jacobian were then defined, two of which are generally applicable. Direct numerical computation of the solution, as described in Section 8.3.1, requires approximately 90 multiplies, whereas the technique described in Section 8.3.3 takes 78 multiplies (Paul, 1981). In comparing the numerical approach of Section 8.3.1 with the differentiation technique of Section 8.3.3, it would appear that they are of comparable complexity, at least if counting multiply operations is indicative. However, it should be noted that the numerical approach also includes a large number of tests and program control overhead which does not occur when using the closed-form solution. Thus, it is highly desirable, if possible, to perform a derivation of the inverse Jacobian in closed form. Furthermore, as noted by (Hollerbach, 1984), *a priori* knowledge of the arm geometry can greatly reduce the complexity in certain cases (e.g., when all 3 wrist axes intersect).

## 8.5 SYNOPSIS

### Vocabulary

You should know the definition and application of the following terms:

Cartesian displacement vector
Cartesian velocity
differential displacement
differential operator
differential rotation
Gaussian elimination
Jacobian
joint velocity
numerical inversion
singularity
small-angle assumption
symbolic inversion

**Notation**

| Symbol | Meaning |
|---|---|
| $\delta_x$ | Differential rotation about the $x$ axis |
| $\delta_y$ | Differential rotation about the $y$ axis |
| $\delta_z$ | Differential rotation about the $z$ axis |
| $dx$ | Differential translation in the $x$ direction |
| $dy$ | Differential translation in the $y$ direction |
| $dz$ | Differential translation in the $z$ direction |
| $dT$ | Differential transform matrix |
| $\Delta$ | Differential operator |
| $^{H}\Delta$ | Differential operator, referenced to the hand frame |
| $\mathbf{D}$ | A 6-vector differential Cartesian displacement |
| $\mathbf{D}_\theta$ | A 6-vector differential displacement, in joint coordinates |
| $\dot{\mathbf{X}}$ | Vector of Cartesian velocities |
| $\dot{\theta}$ | Vector of joint velocities |
| $J$ | Jacobian matrix |

## 8.6 REFERENCES

HOLLERBACH, J., and SAHAR, G. Wrist-partitioned inverse kinematic accelerations and manipulator dynamics. In IEEE International Conference on Robotics, March, 1984, Atlanta, Ga.

PAUL, R. P. Differential kinematic control equations for manipulators. In *Transactions on systems, man, and cybernetics*. 1981(a).

PAUL, R. P. *Robot manipulators*. Cambridge, Mass.: M.I.T. Press, 1981(b).

RENAUD, M. Contribution à la modelisation et à la commande dynamique des robots manipulators. Ph.D. thesis, l'Université Paul Sabatier de Toulouse, 1980.

STEWART, G. W. *Introduction to matrix computations.* New York: Academic Press, 1973.

## 8.7 PROBLEMS

1. Prove the second case of theorem 8.2; that is, a differential rotation about $z$ followed by a differential rotation about $x$ has the same effect as do the two rotations performed in reverse order.

2. Suppose that the hand configuration is given by

$$
T_H = \begin{bmatrix} 0 & 0 & 1 & 1 \\ 1 & 0 & 0 & 6 \\ 0 & 1 & 0 & 3 \\ 0 & 0 & 0 & 1 \end{bmatrix}
$$

What is the effect of a differential rotation of 0.15 rad about the $z$ axis followed by a differential translation of $[0.5, 0.5, 1.0]^T$?

3. Programming assignment: $\sin \theta = \theta$ is an assumption that we believe to be good for small values of $\theta$. Beginning at $\theta = 0$, and increasing in steps of 0.05 rad up to $\pi/2$, plot the percentage error resulting from this assumption.

4. Prove that the Jacobian given in Eq. 8.27 and the inverse given in Eq. 8.31 are in fact inverses of one another.

5. Programming assignment: Assume that the $\theta$-$r$ manipulator moves along the line $y = 1$, from $x = -3$ to $x = 3$, at velocity $\dot{x} = 1$. Compute the radial velocity, $\dot{r}$, and the angular velocity, $\dot{\theta}$, every 0.1 units.

6. Using the strategy described in Section 8.2.3, find $^H J_{21}$. That is, find $\partial y / \partial \theta_1$ for the articulated manipulator.

7. (a) The articulated manipulator is in the configuration given in Example 8.2. Given a differential motion of $[0.1, 0, 0, 0.1, 0.2, 0]^T$, find the differential displacement of the hand frame.
   (b) Find $^H \Delta$ for this configuration.
   (c) Find the motion of the hand, as measured in base coordinates; that is, find $dT_H$.

8. Given that the articulated manipulator is in the following state

$$
^R T_H = \begin{bmatrix} 0 & 1 & 0 & 1 \\ 1 & 0 & 0 & 2 \\ 0 & 0 & -1 & 5 \\ 0 & 0 & 0 & 1 \end{bmatrix}
$$

find the joint velocities corresponding to a Cartesian velocity of

$$
\begin{bmatrix} \dot{x} \\ \dot{y} \\ \dot{z} \\ \dot{\phi}_x \\ \dot{\phi}_y \\ \dot{\phi}_z \end{bmatrix} = \begin{bmatrix} 0.2 \\ 0.2 \\ 0.7 \\ 0 \\ 0 \\ 0.1 \end{bmatrix}
$$

# 9

## PATH CONTROL

In Section 1.4, we described the performance of a typical simple robot using point-to-point control:

Go to point A, stop.
Go to point B, stop.
Open gripper.
Go to point A, stop.
Go to point C, stop.
And so on.

The robot could be permitted to travel between stopping points along any path, usually unconstrained.

Programming a robot by specifying its tasks in terms of a series of points at which it must arrive and stop is not always possible. If the robot were using a paint sprayer, a sander, or an arc welder or were carrying a tray of drinks, then such start-stop motion would be totally unsatisfactory, if not disastrous.

For many applications, it is necessary for the robot's hand to move smoothly in space along a set of points that define its *path* (or *trajectory*).*

In this chapter, we will discuss several approaches to this problem.

---

*In this book we distinguish between the terms "trajectory" and "path." The terms are defined as follows by Brady (1983): "a trajectory is the time sequence of intermediate configurations of the arm between the source $P_0$ and the destination $P_1$ .... The space curve traced by the end effector is called the path of the trajectory," p. 27.

We will discover that path control for a static environment is quite simple but that if the environment is changeable, then path control can become much more complex.

## 9.1 PATH RECORDING

In this section, we will consider the simplest case of path control, in which the robot is led through its motions by a human and simply records the motions and plays them back.

Consider a spray painting application. A robot is rigged with a spray gun. An experienced spray painter then grasps the robot's hand and leads it through a spraying sequence. Each of the robot's joints is equipped with a potentiometer providing direct analog readout of joint position. A multichannel analog tape recorder is running, with one channel connected to each potentiometer. In this way, every motion the operator makes is recorded as a set of instantaneous joint positions (Figure 9.1).

Upon playback, the outputs of the recorder are simply fed to the $\theta_d$ inputs of the joint servos, and the robot will play back the sequence which it was taught.

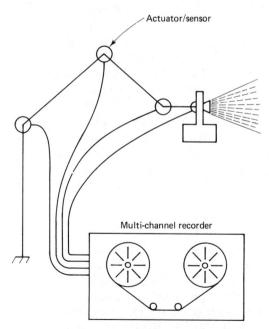

**Figure 9.1**   Use of a multichannel recorder to record joint positions and play them back as set points.

This scheme can be converted to digital form quite readily, by replacing the potentiometers with digital position transducers and using a computer-based data acquisition system, perhaps storing positions on a floppy disk.

The simplicity of such a recording and playback technique makes it very attractive. In addition, we are guaranteed that the resulting path is physically realizable, given that the robot is at least as strong as its trainer.

In many spray painting applications, the parts to be painted are brought to the robot on a conveyor (typically overhead), and the painting is done as the conveyor moves. Accommodation to variations in conveyor speed is reasonably simple. The robot is taught at one speed. The conveyor is instrumented so that its speed controls the speed with which new robot position values are read. In spray painting applications, the spray pressure may need to be adjusted also to avoid coating too thickly or too thinly as the conveyor speed varies.

There are some problems with this technique, of course. The most obvious is the fact that the parts to be painted must be in well-defined positions at all times. This is not so difficult with painting, since the spray area is broad and small positioning errors are insignificant. In arc welding, however, accuracy is critical, and precise fixturing can be excessively expensive or impossible.

Another drawback to the "lead-through" approach is the fact that the robot must be under power during teaching. Robots are massive and difficult to handle and move. During the teach mode, power may be applied to the robot, and gravity compensation set up in the servos. This makes it easy (physically) to program the machine, but it introduces a human into the robot's work volume, a potential hazard.

Nevertheless, simple path recording provides an attractive means for achieving continuous path control in those applications for which it is well suited. For those applications for which it is not well suited, when the robot's environment changes and its path must be recomputed, other, more sophisticated, techniques are required.

## 9.2 CARTESIAN MOTION

Cartesian motion reflects the opposite extreme from the path recording described in the previous section. In Cartesian control, the configuration of the hand is specified in Cartesian coordinates at each point along the path. It is clear that a great many transformations will need to be performed to achieve such control. One can identify two distinct cases of Cartesian control, one in which the calculations are done prior to the motion and a second in which the transforms are performed in real time as the arm is moving.

### 9.2.1 Off-Line Path Computation

In this case, we assume that the path is specified well before the motion is to commence, so that the computer will have adequate time to plan the path in detail. Figure 9.2 shows two alternative ways to represent the path of a manipulator. At each sample point in time, say, every 5 ms, the position may be specified in terms of $(x, y, z, \phi_x, \phi_y, \phi_z)$ or as $(\theta_1, \theta_2, \theta_3, \theta_4, \theta_5, \theta_6)$. In the precomputed case which we are considering now, the two are related by the forward kinematic transform.

The off-line procedure is, thus,

1. Choose the total time for the motion, and divide that time into equal intervals, $\Delta t$.
2. For each time interval, find the configuration of the hand, ${}^R T_H$.
3. Using the forward kinematic transform, find the set of joint positions $\theta_1, \cdots, \theta_6$ corresponding to ${}^R T_H$. Store those positions in a table.

| Time (ms) | Hand position and orientation | | | | | | or Joint angle | | | | | |
|---|---|---|---|---|---|---|---|---|---|---|---|---|
| | X | Y | Z | Roll | Pitch | Yaw | $\theta_1$ | $\theta_2$ | $\theta_3$ | $\theta_4$ | $\theta_5$ | $\theta_6$ |
| 5 | 10 | 2 | 3 | 10° | 20° | 30° | 10° | 35° | 70° | 92° | 18° | 36° |
| 10 | 12 | 2 | 3 | 11° | 20° | 31° | 11° | 34° | 70° | 90° | 18° | 36° |
| ⋮ | | | | | | | | | | | | ⋮ |

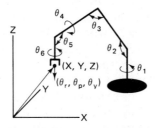

**Figure 9.2** (Snyder 80) Two alternate ways of representing the path of a robot manipulator. This example assumes the robot has only angular joints. For a machine with linear joints, simply replace the $\theta_i$ with $r_i$. (*Courtesy Robotics Age*)

At playback time, the computer simply reads a vector value from the table and specifies that set of joint positions as the $\theta_d$ for each joint. As in the case of path recording, the playback speed can be changed simply by adjusting the clock rate.

Let us look at this problem a little more closely by considering a straight path between two points, stopping at both ends, with no change in orientation. A straight-line path in $x$, $y$, and $z$ will require constant velocity along each axis,

$$\frac{\partial x}{\partial t} = V_x \qquad \frac{\partial y}{\partial t} = V_y \qquad \frac{\partial z}{\partial t} = V_z$$

If the sampling period is $\Delta t$ (say, 5 ms), then the displacement of the hand in the $x$ direction should be $\Delta t V_x$. Thus, for $V_x = 100$ cm/sec (centimeter per second), and $\Delta t = 5$ ms, the table of position values for $x$ should look like

TABLE 9.1    Time-Position Pairs
in a Robot Path

| ms | mm |
|----|----|
| 5  | 5  |
| 10 | 10 |
| 15 | 15 |

The problem with this approach is that it does not take into account accelerations. Table 9.1 requires the arm to go from 0 to 100 cm/sec in 5 ms, an acceleration of 200 m/sec$^2$, or 20 times the acceleration of gravity! It is doubtful that any conventional robot could attain such an acceleration; thus errors would occur in the path. But far worse than the path errors is the fact that the robot would *try* to attain that acceleration. Robots, particularly hydraulic models, are quite powerful and are capable of excessively high accelerations. (There is an apochryphal story of a hydraulic robot which, when presented with a command such as shown for joint 1 in Table 9.1, accelerated at full power, leaving its gripper floating—briefly—in space). It is necessary, therefore, to give some attention to accelerations and decelerations when planning a path.

Having made this observation, it is reasonably straightforward to plan the path in such a way that the hand undergoes a smooth acceleration. We simply plot a straight line in space between the desired position and the current position. We do not, however, divide that path into equal segments to be achieved at equally spaced points in time since we do not require that the robot undergo as rapid a motion at first.

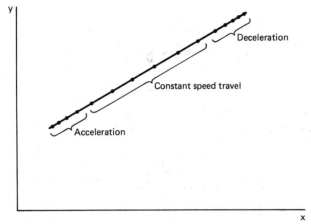

**Figure 9.3**   Trajectory of a robot hand in $x$ and $y$. The set points are indicated by dots along the line. The servos are given new set points at equal time intervals.

In fact, the spacing between is typically parabolic.  Figure 9.3 shows one such path, in $x$ and $y$.  We will examine in more detail a technique for developing a path which takes into account accelerations by using path polynomials.

### 9.2.2 Path Control Polynomials

To accomplish a smooth motion, we need to move the arm in such a way that the hand follows a path which is continuous in position, velocity, and acceleration.  If we define the path of each variable ($x, y$, $z$, etc.) to be a polynomial function of time, we can find coefficients for the polynomial that will make these conditions true.  Before we present the general solution to the design of such polynomials, we will consider a simpler case.

Let $x(t)$ represent the position of the hand in the $x$ coordinate direction.  We wish to accelerate the hand from rest until at time $T$, it has a specified velocity.  Furthermore, at that time, $T$, the hand should be in a specified position, $x = C$, and be moving at a constant velocity, $\dot{x} = C/T$, as shown in Figure 9.4.

In Figure 9.4, note that the motion actually begins at $t = -T$. This choice of time origin makes the mathematics work out much more compactly.

Since we wish to specify six parameters (position, velocity, and acceleration at both ends of the motion), we might choose a fifth-order polynomial.  However, due to the symmetry of the problem, a fourth-order expression is sufficient.  Thus, we represent the path as

$$x(t) = a_4 t^4 + a_3 t^3 + a_2 t^2 + a_1 t + a_0 \qquad (9.1)$$

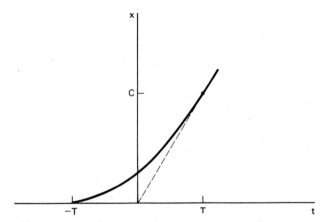

**Figure 9.4**   Path of $x(t)$ so that at time $t = T$, $\dot{x}(T) = C/T$ and $\ddot{x}(T) = 0$.

where the coefficients are unknown and must be found to meet the boundary conditions.

Differentiating Eq. 9.1 with respect to time, we find the expressions for velocity and acceleration:

$$\dot{x}(t) = 4a_4 t^3 + 3a_3 t^2 + 2a_2 t + a_1 \tag{9.2}$$

$$\ddot{x}(t) = 12a_4 t^2 + 6a_3 t + 2a_2 \tag{9.3}$$

From Figure 9.4, we specify boundary conditions as follows:

$$x(-T) = 0 \qquad \dot{x}(-T) = 0 \qquad \ddot{x}(-T) = 0 \tag{9.4}$$

$$x(T) = C \qquad \dot{x}(T) = \frac{C}{T} \qquad \ddot{x}(T) = 0 \tag{9.5}$$

We begin the solutions of this set of equations by evaluating Eq. 9.3 at $t = -T$

$$0 = 12a_4 T^2 - 6a_3 T + 2a_2 \tag{9.6}$$

and at $t = T$,

$$0 = 12a_4 T^2 + 6a_3 T + 2a_2 \tag{9.7}$$

By subtracting Eqs. 9.6 and 9.7, we determine that $a_3 = 0$.

Similarly, the other coefficients can be determined (solution is left as a homework problem), resulting in

$$a_4 = \frac{-C}{16T^4} \qquad a_1 = \frac{C}{2T}$$

$$a_3 = 0 \tag{9.8}$$

$$a_2 = \frac{3C}{8T^2} \qquad a_0 = \frac{3C}{16}$$

**Example 9.1**    Path Control Polynomials

Determine a path control polynomial for a robot to move as follows:

at $t = -3$:

$$x = 0 \qquad \dot{x} = 0 \qquad \ddot{x} = 0$$

$$y = 0 \qquad \dot{y} = 0 \qquad \ddot{y} = 0$$

at $t = 3$:

$$x = 9 \qquad \dot{x} = 3 \qquad \ddot{x} = 0$$

$$y = 15 \qquad \dot{y} = 5 \qquad \ddot{y} = 0$$

**Solution:**    Note that the point $x = 9$, $y = 15$ is a point with slope (relative to the origin) of $15/9 = 5/3$. Furthermore, $dx/dt = 3$ and $dy/dt = 5$ at this point. Therefore,

$$\frac{dy}{dx} = \frac{dy/dt}{dx/dt} = \frac{5}{3}$$

The problem is thus requesting an acceleration from rest onto a straight-line path in $x, y$, with slope $5/3$. This is a common situation.

Evaluating Eq. 9.8, we find a set of numerical values for the coefficients for the path in $x$ and a similar set for $y$. We then have two fourth-order polynomials, one for $x$ as a function of time and one for $y$. In Figure 9.5, we have plotted $y$ versus $x$ for times between $t = -3$ and $t = 3$. The calculations are equally spaced in time, with $\Delta t = 0.2$ sec. We observe straight-line motion connecting smoothly with the constant velocity segment after $t = 3$. Figure 9.5(b) illustrates the position, velocity, and acceleration of the $x$ actuator over the time interval $-3$ to $3$.

**General Form of Path Control Polynomials**

Equations 9.1 through 9.3 can be solved in a more general form, which does not assume zero initial velocity or position (Figure 9.6). After considerable algebraic manipulation, the solution can be written as

$$x(t) = [(V_2 T + \Delta B)(2 - h) h^2 - 2\Delta B] h + \Delta B \qquad (9.9)$$

$$\dot{x}(t) = [V_2 T + \Delta B)(1.5 - h) 2h^2 - \Delta B]/T \qquad (9.10)$$

$$\ddot{x}(t) = (V_2 T + \Delta B)(1 - h) 3h/T^2 \qquad (9.11)$$

$$h = \frac{t + T}{2T} \quad V_2 = \frac{C - B}{T}$$

Note that this solution is to be used between $-T$ and $T$ and that, at time $T$, the axis is moving at a constant speed, which can easily be used for straight-line motion. At time $T$, then, we would switch from following this polynomial simply to tracking along a line of constant velocity.

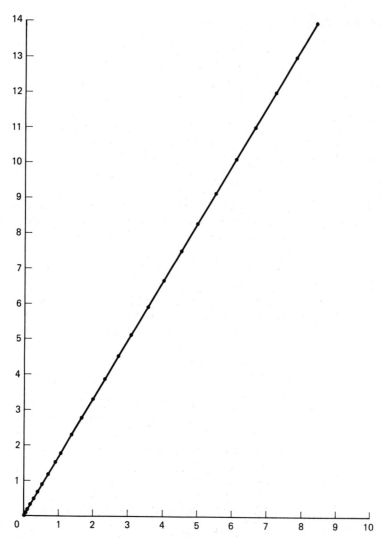

**Figure 9.5(a)**     $y$ versus $x$ for times between $t = -3$ and 3.  The calculations
are equally spaced in time with $x = 0.2$ sec.


**Important Observation**

If, instead of $x$, we used $\theta$ in Eqs. 9.1–9.3, exactly the same mathematics would be applicable, and we would have a technique for finding
motions in which the joint motions would be smooth.  This observation
will be used later in this chapter.

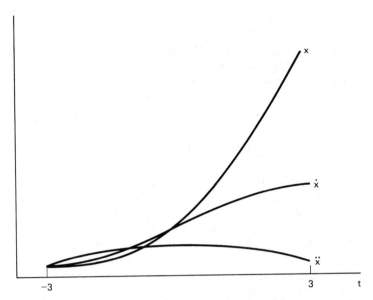

**Figure 9.5(b)**    Position, velocity, and acceleration of the $x$ actuator.

### 9.2.3 Executing a Precomputed Path

In the previous subsection, we developed a method for computing a path that would result in smooth motion. That path was computed (presumably) in Cartesian coordinates. We would then apply the inverse

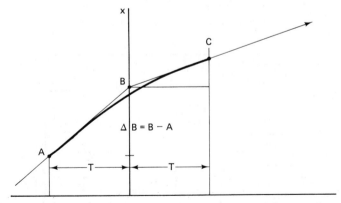

**Figure 9.6**    A path transition from one constant velocity path to another. The joint is moving at velocity $\Delta B/T$ at $t = -T$, when the computer begins to follow the polynomial. At $t = T$, the joint is moving at $(C - B)/T$.

kinematic transform to each point on the path and determine the corresponding joint angles. (Remember, we are still assuming off-line calculations, before motion is to begin, so we have lots of time to do these calculations). Finally, we end up with a table of joint positions, a new position for each sampling instant in time. In this section, we are concerned with the real-time problem of getting the arm actually to follow that path.

### Set Point Tracking (The "Dog Race" Technique)

One way in which to achieve a continuous path is the "carrot and horse" or "dog race" technique. In a dog race, the dogs are released just as the gate is passed by a mechanical rabbit. The rabbit moves at such a speed that the dogs almost, but never quite, catch it.

In the same way, at each instant of time (discrete instants, determined by a clock interrupt), a new value for desired angular position is chosen from the set point table and is used as $\theta_d$ for the joint servo. Figure 9.7 shows this process schematically. The off-line process (Figure 9.7(a)) computes the path and produces as output a table of desired joint positions. In real time, two processes occur concurrently: A servo is running (on each joint), computing joint drive as a function of time and desired joint position. A second process is invoked by a clock

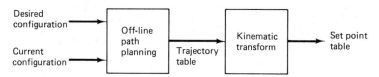

**Figure 9.7(a)**   Off-line path computation.

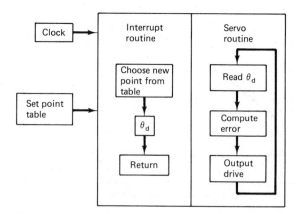

**Figure 9.7(b)**   On-line path control using a set point table.

interrupt.  This process reads a new value from the set point table and stores it in the $\theta_d$ location, effectively changing the servo set point.

Rather than an interrupt process, a second computer may provide the new set point via DMA (direct memory access).  Other, similar variations are possible.

### Velocity-Based Path Control

In place of the dog race technique, another method has been described (Snyder, 1980) that implements the following of a set point table with a minimal requirement for on-line computation.  Figure 9.8 shows the trajectory of a joint as a function of time and a straight-line approximation to this trajectory.  We observe that during each straight-line segment, the (angular) velocity is a constant.  Thus, we may accomplish path control simply by using a velocity servo and commanding a new desired velocity for each segment.  With DC motors, PID (proportional integral derivative) control of velocity can be implemented very simply with a microprocessor of minimal capability (Snyder, 1980). Furthermore, in a hydraulic system, velocity control is simple, since flow is directly controlled and proportional to velocity.

This somewhat naïve approach suffers from the fact that velocities cannot change instantaneously.  Consequently, due to inertia, there is a time lag before the system achieves the desired velocity.  Thus, there is always a positional error.

Figure 9.9 shows a desired joint path (solid line) for a typical motion and the expected response (broken line).  One method to compen-

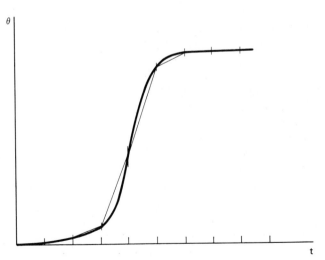

**Figure 9.8**   Trajectory of a joint and straight-line approximation.

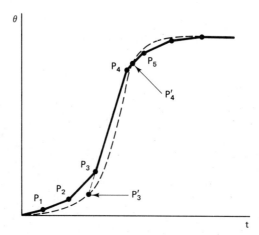

**Figure 9.9**  Desired trajectory (solid line); expected trajectory (broken line).

sate partially for this lag is to command a new velocity earlier than specified. For example, rather than command velocity 4 at time $P_3$ in Figure 9.9, one might command it at time $P_3'$, where the actual trajectory meets the extrapolated next segment.

We can determine when we cross the extrapolated next segment by the following procedure: The equation of a line passing through a given point $(x_1, y_1)$ and having known slope $c$ is

$$y = y_1 + c(x - x_1) \tag{9.12}$$

or

$$y - y_1 - c(x - x_1) = 0 \tag{9.13}$$

To determine whether a given point is above or below the given line, we substitute the $x$ and $y$ values for the point in the LHS (left-hand side) of Eq. 9.13 and test the sign of the result.

If LHS $> 0$, the point is above the line.
If LHS $= 0$, the point is on the line.
If LHS $< 0$, the point is below the line.

To determine the point where the actual trajectory cuts the next segment, we first have to decide the region from which we are approaching the line. Once we know the region of our approach, to find the change point, we just look for the trajectory to move into the opposite region. The region of approach is determined by comparing the present desired velocity with the next desired velocity. This essentially tells us whether we are about to accelerate or decelerate. The two cases are

1. If the next desired velocity is greater than the present desired velocity, we are approaching from a positive region and should look for a negative value of LHS to change the desired velocity.
2. If the desired velocity is smaller than the present desired velocity, we are approaching from a negative region and should look for a positive value of LHS to change the desired velocity.

There are some problems with the criteria just given for changing segments. In segment $P_3$ - $P_4'$, the joint acquires a very high velocity. If we wait until point $P_4'$ to change velocities, we will be guaranteed to overshoot. A simple solution to this problem which yields satisfactory results is given by the following heuristic: Always change the desired velocity at the end of a segment, even if the actual velocity has not reached the next segment yet. On the other hand, if the actual trajectory reaches the next segment earlier than the "end-of-segment" time, the change is made as described earlier. Thus the heuristic dictates that the desired velocity will be changed at the earliest point, either when the schedule specifies or when the actual trajectory reaches the next segment.

Thus, we have developed a technique for following a precomputed path which, by making use of velocity control, requires very little real-time computation. The student should understand this technique, as it exemplifies a number of basic principles. As a practical technique, however, it must be supplemented with conventional PD (proportional derivative) or PID control of positioning to move the joint into precise final position.

### Cartesian Control

The term *Cartesian control* specifically excludes the off-line path calculations described in previous sections and requires, instead, that the calculation of path points, as well as the transformations to joint coordinates, be performed in real time, as the robot is moving. For example, the Cincinnati Milacron $T_3$ robot always moves the hand along straight lines with controlled acceleration and deceleration.

Cartesian control is the most general and most precise technique available. It allows interactions with parts whose position is a function of time and/or an external sensor. Such an application occurs when a robot must work on parts on a moving conveyor. For example, consider a conveyor tracking problem: A robot is to apply a spot weld to a part that is located on a moving conveyor. As the conveyor moves, the part passes a sensor which provides a transform relating the loca-

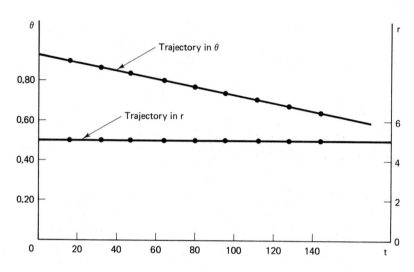

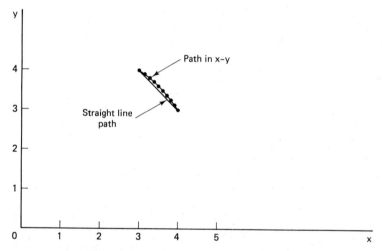

**Figure 9.10**   Results of interpolated motion.

tion of the part to the conveyor:

$$\text{Part relative to conveyor:} \quad {}^{C}T_{P} \tag{9.14}$$

Furthermore, the conveyor is instrumented, so that its position is known at all times relative to the universe frame:

$$\text{Conveyor relative to universe:} \quad {}^{U}T_{C}(t) \tag{9.15}$$

The fact that the conveyor position is a function of time has been made explicit in Eq. 9.15. To place the hand at the appropriate place on the part, we must solve

$$^{U}T_R \cdot {}^{R}T_H = {}^{U}T_C(t) \cdot {}^{C}T_P \qquad (9.16)$$

Then,

$$^{R}T_H = {}^{U}T_R^{-1} \cdot {}^{U}T_C(t) \cdot {}^{C}T_P \qquad (9.17)$$

$$\uparrow \qquad \uparrow \qquad \uparrow$$

$$\text{constant} \quad \text{function} \quad \text{constant}$$
$$\text{of}$$
$$\text{time}$$

Then, as both the conveyor and robot move, Eq. 9.17 must be continuously updated to provide a new desired hand configuration. This requirement is computationally difficult, but possible with modern computers, particularly if arranged in one of the architectures described in Chapter 14. Most often, however, it is used at times when precise positioning is important, but the manipulator is moving slowly, such as when operations must be performed on a part. Approximate methods, described in the next section, can be used to get the robot close to the part at high speeds, with Cartesian control used only for the final, low-speed approach.

## 9.3 JOINT INTERPOLATED CONTROL

There are several control strategies that compromise between simple point-to-point control and the full Cartesian control described in the last section. The objective of these compromises is to provide smooth, accurate path control without the computational burden involved in doing coordinate transforms at very high rates. The reference list for this chapter cites relevant publications. In this section, we describe one such strategy, *joint interpolated control*, which is representative of compromise control techniques.

In developing a strategy for path control, we could define the following desirable characteristics:

1. Continuity of position, velocity, and acceleration
2. Precise control of motion at, or close to the workpiece
3. Midrange motion which is predictable and only stops when necessary

Characteristic 2 can only be achieved in general by using Cartesian control; however, characteristics 1 and 3 are achieved by the following technique.

**Joint Interpolated Motion: Algorithm 1**

1. Divide the path into a number of segments, say, $k$. (We will later discuss the choice of number of segments.)
2. For each segment, transform the Cartesian configuration at each end, and determine starting and ending angular positions for each joint.
3. Determine the time required to traverse each segment.

$$T_{seg} = \max \frac{\theta_{i1} - \theta_{i2}}{\omega_i}$$

where $\omega_i$ is the known (a priori) maximum velocity for joint $i$.
4. Divide $T_{seg}$ into $m$ equal time intervals, $\Delta T_{seg}$, where $m = T_{seg} * f_{samp}$ and $f_{samp}$ is the sampling frequency (typically 50 to 60 Hz). Then, $\Delta T_{seg} = 1/f_{samp}$.
5. For each joint, determine the angular distance to be traveled during $\Delta T_{seg}$:

$$\Delta\theta_i = \frac{\theta_{i1} - \theta_{i2}}{m}$$

6. When motion begins, at the $n$th sampling time, joint servo $i$ receives set point $n\theta_i$.

Using joint interpolated motion, all the joints begin and end their motions simultaneously. Furthermore, the path is smooth and predictable, although not necessarily along a straight line in space. To see this most clearly, consider an example using the $\theta$-$r$ manipulator.

**Example 9.3  Joint Interpolated Motion**

The manipulator is in configuration $(x_1, y_1) = (3, 4)$ and must move to position $(x_2, y_2) = (4, 3)$. Transforming $(x_1, y_1)$ to joint coordinates, we find

$$(r_1, \theta_1) = (5, 0.93) \quad \text{and} \quad (r_2, \theta_2) = (5, 0.64)$$

Hence,

$$\Delta r = 0 \quad \text{and} \quad \Delta\theta = 0.29$$

We know by past experience that the maximum velocities for this manipulator are

$$\dot{r}_{max} = 2 \text{ m/sec}$$

$$\dot{\theta}_{max} = 2 \text{ rad/sec}$$

Then, since $r$ is not required to move,

$$T_{seg} = \frac{0.29 \text{ rad}}{2 \text{ rad/sec}} = 0.15 \text{ sec}$$

If $f_{samp}$ = 60 Hz, then $\Delta T_{seg}$ = $1/f_{samp}$ = 16 ms and $m$ = 0.15 sec/16 × $10^{-3}$ sec = 9. We then find the following points:

| $t$ | $r_d$ | $d$ | $x$ | $y$ |
|-----|-------|------|------|------|
| 0   | 5     | 0.93 | 3    | 4    |
| 16  | 5     | 0.89 | 3.11 | 3.90 |
| 32  | 5     | 0.87 | 3.24 | 3.80 |
| 48  | 5     | 0.84 | 3.36 | 3.70 |
| 64  | 5     | 0.80 | 3.47 | 3.59 |
| 80  | 5     | 0.77 | 3.59 | 3.47 |
| 96  | 5     | 0.74 | 3.70 | 3.36 |
| 112 | 5     | 0.71 | 3.80 | 3.23 |
| 128 | 5     | 0.67 | 3.91 | 3.11 |
| 144 | 5     | 0.64 | 4    | 3    |

As we observe from Figure 9.10, the trajectories in $r$ and $\theta$ are both straight lines; however, the path in $x$-$y$ traced out by the hand is an arc.

Just as path polynomials can be used to provide continuity of acceleration and velocity, they can be used to provide that continuity in joint interpolated motion. Exactly the same techniques are used as were described in Section 9.2.2, but in joint space.

### How Many Segments?

In the previous section, we showed that joint interpolated motion does not, in general, yield a straight-line path. At the end of a segment, the arm is at the correct configuration; therefore, joint interpolated path control may be made more accurate by breaking the path up into more segments, requiring exact positioning at the end of each segment, and using joint interpolated control on each segment. The intersection of two segments is known as a *knot point*. To see the effects of increasing the number of segments, we will consider another example using the $\theta$-$r$ manipulator.

In this case, we will track a conveyor. In Figure 9.11, we show a path being approximated by an arc of radius $R$. $\Delta x$ represents the maximum positive or negative excursion which we will allow from the path. Then $2\Delta y$ is the maximum distance over which a straight-line path be represented by an arc to within the tolerance of $\Delta x$.

Solving the right triangle involving $R$ and $\Delta y$, we have

$$R^2 = \Delta y^2 + h^2 \tag{9.18}$$

but

$$h = R - 2\Delta x \tag{9.19}$$

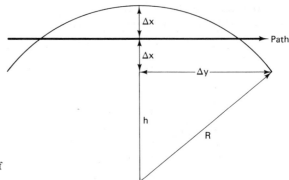

**Figure 9.11** Approximation of
a straight line by an arc.

and

$$R^2 = \Delta y^2 + (R - 2\Delta x)^2 \qquad (9.20)$$

which yields

$$\Delta y^2 = 4R\Delta x - 4\Delta x^2 \qquad (9.21)$$

For a typical robot conveyor tracking task, $R$ might be 1.5 m, and the maximum allowable tracking error could be 1 mm. Then

$$\Delta y^2 = 4(1.5)(10^{-3}) - 4(10^{-3})(10^{-3})$$
$$= 6.0 \times 10^{-3} - 4 \times 10^{-6} \qquad (9.22)$$
$$\cong 6.0 \text{ mm}$$

and

$$2\Delta y = 15.4 \text{ cm}$$

Therefore, in a realistic application, we can sustain 1 mm of accuracy in conveyor tracking by using joint interpolated motion between knot points and placing the knots every 15.4 cm.

Choosing knot points for straight-line motion may be automated by the following recursive procedure.

### Joint Interpolated Motion: Algorithm 2

In this algorithm, we assume we are given initial position and orientation described by the 6-vector $\mathbf{F}_0$ and final position and orientation by the vector $\mathbf{F}_1$. We are also given maximum allowable deviation in position, $\delta_{p\,max}$, and in rotation $\delta_{R\,max}$.

$$\delta_p \leqslant \delta_{p\,max} \qquad \delta_R \leqslant \delta_{R\,max}$$

1. Compute the arm solution for $\mathbf{F}_0$ and $\mathbf{F}_1$, resulting in $\theta_0$ and $\theta_1$, respectively.

2. Compute the joint space midpoint

$$\theta_m = \theta_1 - (\tfrac{1}{2}) \Delta\theta_1$$

   a. From $\theta_m$, use the forward transform to compute $F_m$ = the midpoint of the path that will result from using joint interpolation between $F_0$ and $F_1$.
   b. Compute the desired Cartesian midpoint $F_x = (p_x, R_x)$ where $p_x = (p_0 + p_1)/2$, $R_x = \text{Rot}(n, \psi/2) R_0$. (Any combination of rotations may be represented as a single rotation about a single axis. However, details of finding that axis and angle complicates the discussion unnecessarily at this point.)

3. Compute the deviation between $F_x$, where the hand should be, and $F_m$, where joint interpolation will put the hand.

$$\delta_p = |p_m - p_x|, \; \delta_R = \cos^{-1}(R_x \cdot R_m)$$

4. If $\delta_p < \delta_{p\,max}$ and $\delta_R < \delta_{R\,max}$, then exit. Compute $\theta_x$ corresponding to $F_x$, and apply steps 2–4 recursively to the segments $F_0 \to F_x$ and $F_x \to F_1$.

This algorithm converges very rapidly. The maximum deviation is typically reduced by a factor of four at each level. It is also fully automatic.

## 9.4 CONCLUSION

In this chapter, we have discussed several methods of controlling the motion of a robot's hand over a path in space. The first method discussed was simple playback. The points to be played back could either be simply recorded by monitoring the motion of an operator or computed in an off-line program. In both cases, playback simply consists of constantly reading new set points from a table and passing those points to the servos.

The prospect of real-time control of a Cartesian path introduces not only a great deal of computation, but also the requirement that those computations be done very quickly. In many cases, such control can only be accomplished if the robot is constrained to moving slowly.

There are two fundamental reasons for requiring low speed for Cartesian motion: The first is that the computer must keep up with the arm. The second, as we shall see in the next chapter, is that we need to be able to either compensate for dynamic effects or ignore them. At low speeds, they can be ignored.

The compromise strategies (of which we have discussed only joint interpolated motion) provide reasonably good—almost Cartesian—control, with much lower computational burden.

Some robot languages give the user explicit control over the path control strategy to be used. For example, in VAL, one may specify either

MOV T    or    MOVES T

Both commands move the hand into alignment with a frame named T, but MOVES indicates Cartesian control.

## 9.5 SYNOPSIS

### Vocabulary

You should know the definition and application of the following terms:

acceleration
boundary conditions
joint interpolated control
lead through
path
path control
path control polynomials
playback rate
segment
set point
set point tracking
trajectory
velocity control

### Notation

| Symbol | Meaning |
| --- | --- |
| $\phi_x$ | A component of orientation due to rotation about the $x$ axis (yaw) |
| $\phi_y$ | A component of orientation due to rotation about the $y$ axis (pitch) |
| $\phi_z$ | A component of orientation due to rotation about the $z$ axis (roll) |
| $\theta_i$ | The $i$th joint variable |
| $V_x$ | Velocity in the $x$ direction |
| $V_y$ | Velocity in the $y$ direction |

| Symbol | Meaning |
| --- | --- |
| $V_z$ | Velocity in the $z$ direction |
| $\Delta_t$ | Interval of time between samples |
| $^C T_P$ | Transformation relating a part to a reference frame on a conveyor |
| $^U T_C(t)$ | Transformation relating the conveyor frame to the universe (this transformation normally is a function of time) |
| $T_{seg}$ | Time/segment in joint interpolated control |
| $f_{samp}$ | Sampling frequency |
| $\theta_{i1}$ | Angular position of the $i$th joint at the beginning of a segment |
| $\theta_{i2}$ | Angular position of the $i$th joint at the end of a segment |
| $\omega_i$ | The maximum angular velocity of the $i$th joint |
| $T_{seg}$ | Time required to traverse a segment |
| $\Delta T_{seg}$ | Time required to traverse one part of a segment |
| $\Delta\theta_i = \dfrac{\theta_{i1} - \theta_{i2}}{m}$ | Distance to be traversed during $\Delta T_{seg}$ |

## 9.6 REFERENCES

BRADY, M. Basics of robot motion planning and control. In *Robot motion*, pp. 1–50. M. Brady, J. Hollerbach, T. Johnson, T. Lozano-Pérez, and M. Mason, eds. Cambridge, Mass.: M.I.T. Press, 1983.

COIFFET, P. *Robot technology.* Vol. 1: *Modelling and control.* Englewood Cliffs, NJ.: Prentice-Hall, 1983.

PAUL, R. P. Manipulator cartesian path control. In *IEEE Transactions on Systems, Man, and Cybernetics, 9*, 1979, pp. 702–711. Also in *Robot motion*, planning and control, pp. 245–263. Brady, Hollerbach, Johnson, Lozano-Pérez, and Mason, eds. Cambridge, Mass.: M.I.T. Press, 1983.

SNYDER, W. E. Microprocessor-based path control. *Robotics Age*, 1980.

TAYLOR, R. H. Planning and execution of straight line manipulator trajectories. *IBM Journal of Research and Development, 23*, 1979, pp. 424–436; in *Robot motion, planning and control*, pp. 265–286. Brady, Hollerbach, Johnson, Lozano-Pérez, and Mason, eds. M.I.T. Press, 1983.

WHITNEY, D. E. Resolved motion rate control of manipulators and human prosthesis. *IEEE Transactions on Man-Machine Systems, 10*, 1969.

———. The mathematics of coordinated control of prosthetic arms and manipulators. *Journal of Dynamic Systems, Measurement, and Control* pp. 303–309

(1972); also in *Robot motion, planning, and control*, pp. 287–304. Brady, Holler-bach, Johnson, Lozano-Pérez, and Mason, eds. Cambridge, Mass.: M.I.T. Press, 1983.

## 9.7 PROBLEMS

1. How many 8-bit bytes of memory must be used to store 1 minute of continuous path data for a six-degrees-of-freedom arm using the path recording strategy of Section 9.1? Assume that 16-bit words are required to store each joint position and a sampling interval of 16 ms is used.

2. Equation 9.1 describes the $x$ position of the hand as a function of time. Using the boundary condition specified in Eqs. 9.4 and 9.5, evaluate $a_0$-$a_4$, thus proving the correctness of Eq. 9.8.

3. Programming assignment: Implement Eq. 9.1 in software, using

$$x(t) = a_4 t^4 + a_3 t^3 + a_2 t^2 + a_1 t + a_0$$

and

$$y(t) = b_4 t^4 + b_3 t^3 + b_2 t^2 + b_1 t + b_0$$

with the following boundary conditions:

at $t = -3$:

$$x = 0, \quad \dot{x} = 0, \quad \ddot{x} = 0$$
$$y = 0, \quad \dot{y} = 0, \quad \ddot{y} = 0$$

at $t = 3$:

$$x = 10, \quad \dot{x} = 5, \quad \ddot{x} = 0$$
$$y = 16, \quad \dot{y} = 8, \quad \ddot{y} = 0$$

Plot the results, with 0.2-sec intervals.

4. Suppose that there is a path control system that uses set point tracking with the clock set to provide a new set point every 0.01 sec. The system performs quite well until a hardware failure occurs, such that the interrupts (and hence the new set points) occur every 0.1 sec. What effect will this change will have on the arm performance?

5. Using joint interpolated control of a manipulator with 1 m reach, how many segments are needed to maintain an accuracy of 1 mm while tracking a straight line over a distance of 1 m?

# 10

# *KINETICS*

Sandor (1983) defines dynamics as the analysis of moving bodies. He divides the field of dynamics into two major areas: kinematics, which deals with the geometry and time-dependent aspects of motion without considering the forces causing motion, and kinetics, which is based on kinematics and includes the effects of forces on the motion of masses.

Following this definition, this chapter is entitled *kinetics*, since we now consider masses in motion. The student should be aware, however, that in some places in the literature this same material is referred to as *dynamics*.

Throughout Chapters 6 through 9 we have considered the robot and its constituent parts as three-dimensional geometric structures, but we have ignored the masses and inertias of those structures. We will now introduce this additional complexity to our model of a robot.

In this chapter, we will address several aspects of the issue of kinetics. First, the kinetics of the $\theta$-$r$ manipulator will be derived. We will then discuss the difficulties of solving the kinetics for a general six-degrees-of-freedom manipulator. After we have developed a good understanding of exactly what the kinetic equations *are*, we will consider how they may be used in robotic systems.

After vision, kinetics is probably the most popular research topic in the field of robotics. All the many different approaches being taken by different researchers present massive analytical and computational problems, so that the full complexity of the problem is beyond the scope of this book. Therefore, in this chapter, the emphasis will be on fundamental principles in kinetics, without going into full generality.

The student should learn to relate his or her background in engineering mechanics to the basic issues of robotics. Should it become necessary for the reader to apply these principles to a practical system, the key sources cited in the reference list will allow this material to be extended.

Since our intent in this chapter is to provide background and basic principles, we will not follow the terminology of the robot literature exactly, but will try to explain concepts in basic terms, using the $\theta$-$r$ manipulator as an example.

## 10.1 DERIVING THE KINETICS

In this section, the kinetics are derived for the $\theta$-$r$ manipulator. The derivation is performed using the Lagrangian. No attempt will be made to derive the kinetics for a full six-jointed arm.

### 10.1.1 The Kinetic Model of the $\theta$-$r$ Manipulator

Figure 10.1 shows the $\theta$-$r$ manipulator, but now we are including loads. Figure 10.2 shows the model schematically. The mass of the outer cylinder is assumed at its center of mass, $m_1$, at *constant* distance $r_1$ from the center of rotation. The telescoping radial arm and load are modeled as a mass $m_2$ at distance $r$.

### 10.1.2 Lagrangian Kinetics

The *Lagrangian* of a mechanical system is defined by

$$L = K - P \tag{10.1}$$

where $K$ is the total kinetic energy of the system and $P$ is the total potential energy.

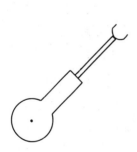

**Figure 10.1** $\theta$-$r$ manipulator.

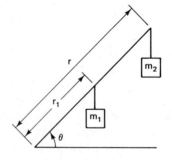

**Figure 10.2** Schematic representation, showing equivalent masses.

If an actuator is controlling a rotary variable, $\theta$, then the torque seen by that actuator is

$$T = \frac{\partial}{\partial t} \frac{\partial L}{\partial \dot{\theta}} - \frac{\partial L}{\partial \theta} \tag{10.2}$$

Similarly, if the joint is prismatic, then the force applied in the direction of motion, $x$, is

$$F_x = \frac{\partial}{\partial t} \frac{\partial L}{\partial \dot{x}} - \frac{\partial L}{\partial x} \tag{10.3}$$

We note that Eqs. 10.2 and 10.3 are identical and could be defined in generalized coordinates.

We begin deriving the kinetics by finding the kinetic energy of the two masses. Considering mass $m_1$ first, its Cartesian position is

$$x_1 = r_1 \cos(\theta) \tag{10.4}$$

$$y_1 = r_1 \sin(\theta) \tag{10.5}$$

We differentiate with respect to time to get Cartesian velocities, noting that $r_1$ is a constant.

$$\dot{x}_1 = -r_1 \sin\theta\dot{\theta} \tag{10.6}$$

$$\dot{y}_1 = r_1 \cos\theta\dot{\theta} \tag{10.7}$$

Then, the magnitude of the velocity vector is

$$V_1^2 = (\dot{x}_1)^2 + (\dot{y}_1)^2 \tag{10.8a}$$

$$\begin{aligned} V_1^2 &= r_1^2 \sin^2\theta\dot{\theta}^2 + r_1^2 \cos^2\theta\dot{\theta}^2 \\ &= r_1^2 \dot{\theta}^2(\sin^2\theta + \cos^2\theta) \end{aligned} \tag{10.8b}$$

$$V_1^2 = r_1^2 \dot{\theta}^2 \tag{10.9}$$

The kinetic energy of a mass $m$ moving at velocity $V^2$ is

$$K = \tfrac{1}{2} mV^2 \tag{10.10}$$

so

$$K_1 = \tfrac{1}{2} m_1(r_1^2 \dot{\theta}^2) \tag{10.11}$$

The derivation for $K_2$ is similar, complicated only be the fact that $r$ is not a constant.

$$x_2 = r \cos\theta \tag{10.12}$$

$$y_2 = r \sin\theta \tag{10.13}$$

$$\dot{x}_2 = \dot{r} \cos\theta - r\dot{\theta} \sin\theta \tag{10.14}$$

$$\dot{y}_2 = \dot{r} \sin \theta + r\dot{\theta} \cos \theta \tag{10.15}$$

$$V_2^2 = (\dot{r} \cos \theta - r\theta \sin \theta)^2 + (\dot{r} \sin \theta + r\dot{\theta} \cos \theta)^2 \tag{10.16}$$

which simplifies to

$$V_2^2 = \dot{r}^2 + r^2 \dot{\theta}^2 \tag{10.17}$$

Therefore,

$$K_2 = \tfrac{1}{2} m_2 [\dot{r}^2 + r^2 \dot{\theta}^2] \tag{10.18}$$

Now, we find the potential energy, using

$$P = mgh \tag{10.19}$$

where $h$ is the height and $g$ the acceleration of gravity

$$P_1 = m_1 g r_1 \sin \theta \tag{10.20}$$

$$P_2 = m_2 g r \sin \theta \tag{10.21}$$

The total kinetic energy is thus

$$K = K_1 + K_2 = \tfrac{1}{2} m_1 r_1^2 \dot{\theta}^2 + \tfrac{1}{2} m_2 \dot{r}^2 + \tfrac{1}{2} m_2 r^2 \dot{\theta}^2 \tag{10.22}$$

and the total potential energy is

$$P = m_1 g r_1 \sin \theta + m_2 g r \sin \theta \tag{10.23}$$

We combine Eqs. 10.22 and 10.23 to produce the Lagrangian for this manipulator

$$L = \tfrac{1}{2} m_1 r_1^2 \dot{\theta}^2 + \tfrac{1}{2} m_2 \dot{r}^2 + \tfrac{1}{2} m_2 r^2 \dot{\theta}^2 - m_1 g r_1 \sin \theta - m_2 g r \sin \theta \tag{10.24}$$

We will first find the torque about $\theta$ using Eq. 10.2:

$$\frac{\partial L}{\partial \dot{\theta}} = m_1 r_1^2 \dot{\theta} + m_2 r^2 \dot{\theta} \tag{10.25}$$

$$\frac{\partial}{\partial t} \frac{\partial L}{\partial \dot{\theta}} = m_1 r_1^2 \ddot{\theta} + m_2 r^2 \ddot{\theta} + 2 m_2 r \dot{\theta} \dot{r} \tag{10.26}$$

$$\frac{\partial L}{\partial \theta} = -g \cos \theta \, (m_1 r_1 + m_2 r) \tag{10.27}$$

$$T_\theta = m_1 r_1^2 \ddot{\theta} + m_2 r^2 \ddot{\theta} + 2 m_2 r \dot{r} \dot{\theta} + g \cos \theta \, (m_1 r_1 + m_2 r) \tag{10.28}$$

Having found the torque about the rotational actuator, we are now interested in the force applied by the linear actuator

$$\frac{\partial L}{\partial \dot{r}} = m_2 \dot{r} \qquad (10.29)$$

$$\frac{\partial}{\partial t} \frac{\partial L}{\partial \dot{r}} = m_2 \ddot{r} \qquad (10.30)$$

$$\frac{\partial L}{\partial r} = m_2 r \dot{\theta}^2 - m_2 g \sin \theta \qquad (10.31)$$

$$F_r = m_2 \ddot{r} - m_2 r \dot{\theta}^2 + m_2 g \sin \theta \qquad (10.32)$$

Equations 10.28 and 10.32 together provide a complete description of the kinetics of this manipulator; that is, they provide a relationship between the torque or force applied by an actuator and the resulting motions. The accuracy of the kinetics is largely dependent on the accuracy of the model. For example, Figure 10.2 ignores the effects of friction, and friction may be a major sink for energy.

### The Newton-Euler Approach

The derivation given in the previous section used the Lagrangian to derive the kinetics. The Lagrangian itself can be derived from the fundamental laws of motion:

$$\text{Newton's equation: } \mathbf{F} = m\ddot{\mathbf{x}}$$

and

$$\text{Euler's equation: } \mathbf{T} = \dot{\mathbf{H}}$$

where $\dot{\mathbf{H}}$ is the time rate of change of moment of momentum. Several researchers (Luh, Walker, and Paul, 1983) have derived the kinetics of manipulators using the laws of motion directly. The derivation resulting from the *Newton-Euler approach* is equivalent to the derivation using the Lagrangian. However, the *forms* of the solution are different, and Lee (1983) points out that the form of the Newton-Euler solution requires less computation than does the equivalent Lagrangian solution. However, Hollerbach (1980) points out that

It appears unlikely that any but very minor improvements in efficiency can be made to the present recursive formulations. The recursive Newton-Euler formulation is more efficient than the recursive Lagrangian . . . (however) . . . the Lagrangian formulation is not so computationally intensive as to preclude real time computation, as had been the assumption for the past 10 years (Kahn, 1969; Paul, 1972; Bejczy, 1974; Raibert, and Horn, 1978), and the Lagrangian formulation can be made roughly as efficient as the Newton-Euler formulation (Luh, Walker, and Paul 1980). For some applications involving the use of homogeneous coordinates, . . . the recursive Lagrangian formulation may be the most convenient efficient dynamics formulation. (pp. 84)

The student is encouraged to use the Newton-Euler approach to solve the $\theta$-$r$ manipulator. Note, however, that the center of mass is constantly changing.

## 10.2 FORCES AND TORQUES

In this section we will review the results of the previous two sections and attempt to provide a physical interpretation of the results.

We wish to reorder the terms of Eq. 10.28 so that they have a particular form; specifically, we wish to write

$$
\begin{aligned}
T_\theta &= D_{11}\ddot{\theta} + D_{12}\ddot{r} && \text{(inertial term)} \\
&+ D_{111}\dot{\theta}^2 + D_{122}\dot{r}^2 && \text{(centripetal term)} \\
&+ D_{112}\dot{\theta}\dot{r} + D_{121}\dot{r}\dot{\theta} && \text{(Coriolis term)} \\
&+ D_1 && \text{(gravity term)}
\end{aligned} \tag{10.33}
$$

Putting Eq. 10.28 into this form, we have

$$
\begin{aligned}
T_\theta &= [\,m_1 r_1^2 + m_2 r^2\,]\,\ddot{\theta} + [0]\,\ddot{r} \\
&+ [0]\,\dot{\theta}^2 + [0]\,\dot{r}^2 \\
&+ [\,m_2 r\,]\,\dot{\theta}\dot{r} + [\,m_2 r\,]\,\dot{r}\,\dot{\theta} \\
&+ [\,g\cos\theta\,(m_1 r_1 + m_2 r)\,]
\end{aligned}
$$

Thus,

$$
\begin{array}{ll}
D_{11} = m_1 r_1^2 + m_2 r^2 & D_{12} = 0 \\
D_{111} = 0 & D_{122} = 0 \\
D_{112} = m_2 r & D_{121} = m_2 r \\
D_1 = g\cos\theta\,(m_1 r_1 + m_2 r) &
\end{array}
$$

We will define the prismatic actuator to be actuator 2 (the first subscript of the $D$'s refers to the actuator) and perform a similar reorganization of terms on Eq. 10.32, yielding

$$
\begin{array}{ll}
D_{21} = 0 & D_{22} = m_2 \\
D_{211} = -m_2 r & D_{222} = 0 \\
D_{212} = 0 & D_{221} = 0 \\
D_2 = m_2 g \sin\theta &
\end{array}
$$

Recall from Chapter 5 that the inertia of a point mass at distance $r$ is $mr^2$. Then the term $D_{11}$ can be seen as the *effective inertia*.

Similarly, the term $D_{22}$ acts as a purely inertial load on joint 2. Since joint 2 is prismatic, an inertial load appears as a mass. Thus, if there were no other terms, $D_{11}$ would provide Euler's equation for joint 1 and $D_{22}$ would provide Newton's equation for joint 2.

Even if there were no other terms, we would still have a nonlinear system, since $D_{11}$ varies with the square of $r$ (*only* $r$; recall that $r_1$ is constant).

But there are other terms. $D_{ij}$ for $i \neq j$ does not occur in this arm, but it does in more complex arms. It is known as a coupling inertia. When such terms exist, they indicate that an acceleration of one joint causes a torque at another joint. This is most noticeable in shoulder-elbow interactions.

A term of the form $D_{ijj}\,\dot{\theta}_j^2$ is referred to as the centripetal force acting on joint $i$ as a result of a velocity at joint $j$. In this manipulator, joint 2 is affected by the velocity at joint 1, resulting in a force acting in the radial direction and proportional to $r$.

Terms of the form $D_{ijk}$ and $D_{ikj}$ really need to be treated together, since they represent forces at actuator $i$ due to combined motions of actuators $j$ and $k$. These are known as Coriolis forces.

It is instructive to examine $D_{112}$ more closely. $D_{112}$ contributes to the rotary torque by adding an additional torque term, $D_{112}\,\dot{r}\dot{\theta} = m_2 r \dot{r} \dot{\theta}$. We observe that if mass $m_2$ is not moving radially ($\dot{r} = 0$), no torque occurs. Furthermore, the faster $r$ moves, the greater the torque felt by joint 1. A familiar example of this phenomenon is experienced on a children's playground carrousel. As an experiment, stand on one of these carrousels while it is spinning rapidly. Kick your leg toward the inside. You will find your leg deflected unexpectedly to the left or right (in the radial direction) according to which way the carrousel is spinning. This is the effect of the Coriolis forces.

We have now derived equations that relate the torques and forces on robot actuators to the motions caused by these actuators. In the next section, we will see how one might make use of this information.

### Example 10.1 Numerical Evaluation

The $\theta$-$r$ manipulator is assumed to weigh 10 kg all centered at $m_1$. $r_1 = 1$ m and the load may vary, between 1 and 5 kg. $r$ may range from 1 to 2 m. Maximum velocities are $\dot{\theta} = 1$ rad/sec and $\dot{r} = 1$ m/sec. Maximum accelerations are $\ddot{\theta} = 1$ rad/sec$^2$ and $\ddot{r} = 1$ m/sec$^2$.

Evaluate $T$ in the following conditions:

1. Arm horizontal and fully extended, but stationary
2. Arm horizontal and fully extended, but moving at maximum (constant) velocities in $r$ and $\theta$
3. Arm horizontal and fully extended, stationary, but undergoing maximum radial acceleration

**Solution:**

Case 1. Arm horizontal and fully extended, but stationary. In this case we have only gravity loading of

$$T_\theta = D_1 = (m_1 r_1 + m_2 r) g \cos \theta = (20)(9.8) = 196 \, (\text{kg} - \text{m}^2/\text{sec}^2)$$

Case 2. Arm horizontal and fully extended, but moving at maximum (constant) velocities in $r$ and $\theta$.

$$T_\theta = D_1 + D_{112}\dot\theta\dot r + D_{121}\dot r\dot\theta$$

$$= D_1 + 2m_2 r(\dot\theta\dot r) = D_1 + (2)(5)(2)(1)(1) = 216 \, (\text{kg} - \text{m}^2/\text{sec}^2)$$

We note that even at maximum velocities, the Coriolis forces in this configuration are insignificant compared with the gravity loading.

Case 3. Arm horizontal and fully extended, stationary, but undergoing maximum radial acceleration.

$$T_\theta = D_1 + D_{11}\ddot\theta$$

$$= D_1 + (m_1 r_1^2 + m_2 r^2)(1) = D_1 + 20$$

$$= 212 \, (\text{kg} - \text{m}^2/\text{sec}^2)$$

Again, the gravity loading far exceeds any of the other loads on the arm. Furthermore, the variations in gravity loading are tremendous, going from 192 kg - m$^2$/sec$^2$ in the horizontal configuration to zero in the vertical. We also note that both centripetal and Coriolis forces vary as the product of two velocities and, therefore, approach zero very rapidly as the arm slows.

In general, experience has shown that for most arms, a model of the kinetics which includes only the inertial and gravity terms is quite satisfactory. See Bejczy (1979) or Paul (1981) for more detail.

## 10.3 THE COMPLEXITY OF THE SOLUTION

For a fully articulated arm, the computation of the kinetics becomes quite complex. The equation can be expressed in matrix form as

$$\left\{ \sum_{\substack{j=1 \\ i=1, N}}^{N} \left\{ A(i,j)\,\ddot\theta_j + B(i,j)\,\dot\theta_j^2 + \sum_{k=j+1}^{N} C(i,j,k)\,\dot\theta_j\dot\theta_k \right\} \right\} = Q_i + \Gamma_{\theta_i}$$

or as

$$[A] \quad \ddot\theta \quad + \quad [B] \quad \dot\theta^2 \quad + \quad [C] \quad \dot\theta\dot\theta \quad = Q(\theta) \quad + \quad \Gamma_\theta$$

$$N \times N \quad N \times 1 \qquad N \times N \quad N \times 1 \qquad N \times C_N^2 \quad C_N^2 \times 1 \qquad N \times 1 \qquad N \times 1$$

$$c_{54} = M_5 \left( -C5.\ell_{Y_4}.d_5 - SC5.d_5^2 + SC5.I_{Z_5} \right)$$

$$a_{11} = I_{Z_1} + M_2 SS2\, d_2^2 + SS2.I_{Y_2} + CC2.I_{Z_2} + M_3\, [SS2\, \ell_2^2$$
$$+ 2S2S\, (2+3)\, C_2\, d_3 + SS\, (2+3)d_3\,] + SS(2+3)\, I_{Y_3}$$
$$+ CC(2+3)\, I_{Z_3} + M_4\,[SS2\, \ell_2^2 + 2S2S(2+3)\, C_2\, h_1$$
$$+ 2S2C(2+3)\, C4\, C_2\, d_{Y_4} + SS(2+3)h_1^2 + 2SC(2+3)\, C4\, d_{Y_4}\, h_1$$
$$+ CC(2+3)\, d_{Y_4}^2 + SS\, (2+3)\, SS4 d_{Y_4}^2\,] + SS(2+3)\, SS4\, I_{X_4}$$
$$+ SS(2+3)\, CC4.I_{Y_4} - 2SC(2+3)\, C4\, I_{Y Z_4} + CC(2+3)\, I_{Z_4}$$
$$+ M_5\,[SS2\ell_2^2 + 2 S2 S(2+3)\, C_2\, h_2 + 2S2C\, (2+3)\, C4\, C_2\, C_{Y_4}$$
$$+ 2\, S2\, S(2+3)\, C5\, C_2\, d_5 + 2S2C(2+3)\, C4S5\, C_2\, d_5 + SS(2+3)h_2^2$$
$$+ 2SC(2+3)\, C4\, C_{Y_4}\, h_2 + 2SS(2+3)\, C5\, h_2\, d_5 + 2SC(2+3)\, C4S5\, h_2\, d_5$$
$$+ CC(2+3)\, \ell_{Y_4}^3 + 2SC(2+3)C4C5\, \ell_{Y_4}\, d_5 + 2CC(2+3)\, S5\, \ell_{Y_4} d_5$$
$$+ SS(2+3)CC4CC5\, d_5 + 2SC(2+3)\, C4SC5\, d_5^2 + CC(2+3)\, SS5\, d_5^2\,]$$
$$+ SS(2+3)\, SS4\, C_{Y_4}^2 + 2SS\, (2+3)\, SS4S5\, \ell_{Y_4}\, d_5 + SS(2+3)\, SS4\, d_5^2$$
$$+ SS(2+3)\, SS4\, I_{X_5} + SS(2+3)\, CC4CC5\, I_{Y_5} + 2SC(2+3)\, C4SC5\, I_{Y_5}$$
$$+ CC(2+3)\, SS5\, I_{Y_5} + SS(2+3)\, CC4SS5\, I_{Z_5} - 2SC(2+3)\, C4SC5\, I_{Z_5}$$
$$+ CC(2+3)\, CC5\, I_{Z_5} + M_6\,[SS2\, \ell_g^2 + 2S3S\, (2+3)\, \ell_g\, d_6$$
$$+ SS(2+3)d_6^2\,] + SS(2+3)\, I_{Y_6} + CC(2+3)\, I_{Z_6}$$

**Figure 10.3**    (Coiffet)

Now the matrix $A$ represents the inertial forces, $B$ the coefficient matrix of the centripetal forces, $C$ the coefficient matrix of Coriolis forces, $Q$ the vector of gravitational forces, and $\Gamma_\theta$ the vector of externally applied forces.

To obtain some idea of the complexity of the solution to a typical arm with six degrees of freedom, see Figure 10.3 which shows two typical coefficients. In this particular derivation (Coiffet, 1983), $c_{54}$ was the least complex coefficient and $a_{11}$, the most complex.

Computations of such functions in real time would be difficult to say the least. Therefore, many and varied simplifications have been made in attempts to reduce the complexity to a workable level. These will be discussed in the next section.

## 10.4 USING KINETIC EQUATIONS

In the previous section, expressions were derived that relate the torques and forces exerted by the actuators to the resultant motions. Those derivations were performed without thought of how to make use of such expressions, once derived. In this section, we will discuss several of the ways in which we might use this information.

### 10.4.1 Design of Manipulators

Certainly the first use of the kinetic model is in the design of a manipulator. The designer can enter the geometry of his proposed design along with estimates of link masses, loads, and actuator characteristics and simulate the performance of the arm. Several such simulation packages are available (Nevins et al., 1972; Dillion, 1974; Khalil, 1978). Results of such simulations may be used to choose actuators of the appropriate size, for the dynamic equations tell us precisely the torque required to accomplish any specified motion. The simulations may also be used to indicate required redesign of the mechanical structures.

### 10.4.2 Workcell Design and Path Planning

The path control techniques of Chapter 9 allow the user to specify a desired path for the robot; however, as speeds and acceleration are increased, kinetic effects may result in unexpected deviations from the planned path. Simulations of path control that take into account the kinetic model may be used to develop worst case estimates of path deviations at high speeds. This information can be used in turn to plan the locations of machines and fixtures in a workcell to guarantee correct obstacle avoidance.

### 10.4.3 Using Kinetics in Real-Time Control

As was shown in Chapter 5, no single choice of servo gains is appropriate to provide the best performance of a robot. With the use of a dynamic model of the arm, however, we now have the potential for attaining such optimal control, since we can now describe the interaction of the joints.

In this section, we will mention several strategies which have been used in attempts to attain optimal or at least better dynamic control of manipulators. The complexity of the solution prohibits evaluation of the kinetics at the servo rate. Therefore, virtually every approach we will describe makes use of some simplifying assumptions.

#### Multivariable Control

The control of any system can be put in the form

$$\dot{x}(t) = A(x, t) + B(x, t)\, u(t)$$

where $x$ is a vector of position and $u$ is a vector of control values, and $A$ and $B$ are matrices in which the elements are nonlinear functions of the state of the system and can be derived from the kinetics. Thus, given a desired path or objective, $x(t)$, this expression can be solved (in principle) to obtain a control function $u(t)$ which will accomplish that motion. In order for $u(t)$ to be unique, some other criterion is usually added, such as minimum energy:

$$\int_{motion} u(t)^2 \, dt \text{ is minimal}$$

or others. If we are fortunate, the resulting control can be formulated as a feedback function; that is,

$$\mathbf{u}(t) = f(\mathbf{x}_{desired} - \mathbf{x}(t))$$

for such controls are much more "robust," that is, able to perform well in the presence of inaccuracies in the model (see Gruver, Hedges, and Snyder, 1980).

Such optimal control strategies are the topic of current research in control theory and are not employed in commercial robots at this time. See Mesarovic et al. (1970), Bryson and Ho (1969) or Gruver et al. (1980) for more details.

### Gravity Compensation

In Chapter 5, we noted that an integrating controller achieved gravity compensation by allowing the error signal to build up in an integrator until the torque due to this integrated error exactly canceled the gravity-induced torque. Improperly designed integrating controllers have some lag and a tendency to overshoot. Somewhat better performance can be obtained by computing the $D_i$ term of the kinetic equation and using it in a feedforward loop. That is, the torque out is equal to the torque specified by the servo plus the gravity term, thus canceling the gravity effects.

## 10.5 CHOOSING SERVO GAINS

The dynamic model of the arm may be used to adjust the gains of the servos in such a way as to achieve improved performance. In so doing, one is still faced with computational delays which can lead to instabilities, as well as the fact that models are always inaccurate. For these reasons, methods have been developed which are both self-correcting (robust) and make use of simplified models.

### Linearization

Kahn (1969) developed a system which made use of a dynamic model linearized about an operating point, and using "bang-bang" control. In practice, the linearization proved to be relatively inaccurate.

### Diagonalization

Paul (1972) diagonalized the $A$ matrix. That is, in his model, all elements off the diagonal are set to zero. In Paul (1981), a more thorough derivation is given, allowing limited coupling. Bejczy (1979) also gave careful attention to the relative magnitude of the coefficients and selectively discarded smaller coefficients. In Bejczy's control system, a second (constant gain) feedback loop corrected for positional and velocity errors that might result from control based on an inaccurate model.

### Precomputing Coefficients

Since it is so difficult to calculate the dynamic coefficients in real time, they might be calculated in advance for all possible configurations. However, such a strategy seems hardly feasible, since the required amount of such data would be huge. Raibert (1978) suggests that the configuration space be divided into zones and that the coefficients be assumed constant within a zone.

One could easily take this strategy a step farther and then precompute the appropriate servo gains for each zone. This strategy is sometimes used in commercial controllers.

### 10.6 CONCLUSION

In this chapter, we have discussed the dynamic response of a manipulator. That is, how the motions of the various joints are related to the applied torques and loading of the arm. We discovered that a complete model of an arm results in six coupled second-order nonlinear differential equations. Just deriving those equations is a formidable task, and their analytic solution is impossible.

The dynamic model can, however, provide a valuable tool for the robot or workcell designer as he or she can simulate the performance of hypothetical arms under varying conditions.

The gravity loading terms from the kinetic model can be used in real-time control to improve performance by gravity compensation.

In practice, most systems simply use different servo gains for different configurations; however, the area of robot kinetics is an active research area, and more sophisticated techniques are rapidly moving from laboratories to applications.

## 10.7 SYNOPSIS

### Vocabulary

You should know the definition and application of the following terms:

actuator sizing
centripetal forces
Coriolis forces
coupling forces
diagonalization
Euler's equation
gravity compensation
inertial forces
kinetic energy
Lagrangian
minimal energy control
Newton's equation
potential energy
robust
workcell design

### Notation

| Symbol | Meaning |
| --- | --- |
| $\theta$ | Joint 1 of the $\theta$-$r$ manipulator |
| $r$ | Joint 2 of the $\theta$-$r$ manipulator |
| $m$ | A mass |
| $T$ | An arbitrary torque |
| $T_\theta$ | Torque about the $\theta$ axis |
| $L$ | The Lagrangian |
| $F_x$ | Force in the $x$ direction |
| $x_i, y_i$ | The position of mass $i$ |
| $V_i$ | The velocity of mass $i$ |
| $K_i$ | Kinetic energy of mass $i$ |
| $P_i$ | Potential energy of mass $i$ |
| $g$ | Acceleration of gravity |
| $H$ | Moment of inertia |

| Symbol | Meaning |
|---|---|
| $D_{ii}$ | Inertial coefficient for joint $i$ |
| $D_{ijj}$ | Centripetal coefficient for joint $i$ |
| $D_{ijk}$ | Coriolis coefficient for joint $i$ |
| $D_i$ | Gravity coefficient for joint $i$ |
| $A$ | Matrix of inertial coefficient |
| $B$ | Matrix of centripetal coefficient |
| $C$ | Matrix of Coriolis coefficient |
| $Q$ | Vector of gravitational forces |
| $\Gamma$ | Vector of externally applied forces |
| $\mathbf{u}(t)$ | Vector of control values |

## 10.8 REFERENCES

BEJCZY, A. K. Robot arm dynamics and control. *Jet Propulsion Lab* (TM 33-669), November 1979.

BRYSON, A. and HO, Y. *Applied optimal control.* Waltham, Mass: Blaisdell, 1969.

COIFFET, P. *Robot technology.* Vol. 1: *Modelling and control.* Englewood Cliffs, N.J.: Prentice-Hall, 1983.

DILLON, S. R. Automatic equation generation and its application to problems in control. *Proceedings of Joint Automatic Control Conference,* Austin, June 1974.

GRUVER, W. A., HEDGES, J. C., and SNYDER, W. E. Decentralized optimal control and coordination of large-scale linear systems. In *Optimal control applications and methods,* Vol 1, New York: John Wiley and Sons, February 1980.

HAMAMI, H., JASWA, V., and MCGHEE, R. Some alternative formulations of manipulator dynamics for computer simulation studies. 13th Allerton Conference on Circuit and System Theory. University of Illinois, Urbana, October 1975.

HOLLERBACH, J. M. A recursive Lagrangian formulation of manipulator dynamics and a comparative study of dynamics formulation complexity. *IEEE Transactions on Systems, Man, and Cybernetics, 10,* November 1980. Also in *Robot motion, planning and control,* (pp. 89–106) Brady, M., et al., eds., Cambridge, Mass.: M.I.T Press, 1983.

KAHN, M. E. The near minimum time control of open-loop articulated kinematic chains. Stanford Artificial Intelligence Lab. (AIM 106), December 1969.

KHALIL, W. Contribution to automatic control of manipulators, using a mathematical model of the mechanisms. Thesis, Montpelier, France, 1978.

LEE, C. S. On the control of robot manipulator. *Proceedings of the 27th SPIE Conference,* San Diego, August 1983.

LUH, J. Y. S., WALKER, M. W., and PAUL, R. P. On-line computational scheme for mechanical manipulators. *Journal of Dynamic Systems, Measurement, and Control, 102,* 1980. Also in *Robot motion, planning and control,* pp. 89–106. Brady, M., Hollerbach, J., Johnson, T., Lozano-Pérez, T., and Mason, M., eds., Cambridge, Mass.: M.I.T. Press, 1983.

MESAROVIC, M. D., MACKO, D., and TAKAHARA, Y. *Theory of hierarchical multilevel control.* New York: Academic Press, 1970.

NEVINS, T. L., SHERIDAN, T. B., WHITNEY, D. E., and WOODLIN, A. E. The multimoded remote manipulator system. *Teleoperator Arm Design Report,* E2720 (M.I.T), October 1972.

PAUL, R. P. Modelling, trajectory calculation, and servoing of a computer controlled arm. Stanford Artificial Intelligence Lab. (AIM 177), November 1972.

———. Manipulator Cartesian path control. *IEEE Transactions on Systems, Man, and Cybernetics, 9,* 1979.

PAUL, R. *Robot Manipulators,* Cambridge, Mass.: M.I.T. Press, 1981.

RAIBERT, M. H., and HORN, B. K. P., Manipulator control using the configuration space method, *Industrial Robot, 5,* June 1978.

SANDOR, B. I. *Engineering mechanics: Dynamics.* Englewood Cliffs, N.J.: Prentice-Hall, 1983.

SARIDIS, G. *Advances in automation and robotics: Theory and applications,* Vol 1. Greenwich, Conn.: JAI Press, 1984.

## 10.9 PROBLEMS

1. Programming assignment: Simulate the $\theta$-$r$ manipulator as it moves from $\theta = 0$ to $\theta = 90°$. Assume a constant angular velocity of $\dot{\theta} = 2$ rad/sec and constant radial velocity of $\dot{r} = 0.5$ m/sec. The initial value of $r = 1$ m; $r_1 = 1$ m; $m_1 = m_2 = 1$kg. Every $5°$, compute the following values:
   (a) Inertial coefficient (e.g., $D_{11}$)
       Inertial torque (e.g., $D_{11}\ddot{\theta}$) and forces
   (b) Centripetal coefficients
       Centripetal torques and forces
   (c) Coriolis coefficients, torques, and forces
   (d) Gravity coefficients, torques, and forces

What is the maximum percentage variation in any one term?

2. Rework problem 1, but ignore all terms except inertia and gravity. What is the worst case error that this assumption causes in the computed torque and force?

# 11

# FORCE CONTROL AND COMPLIANCE

In this chapter, we will consider yet another form of control, control of the forces exerted by the robot's hand.

We will be relating forces in Cartesian space to torques in joint space, as we did in Chapter 10. However, in this chapter, we will be dealing with constrained motion, for example, the hand pressing against a table top. Hence, velocities will be very low, and the dynamic effects described in Chapter 10 will not be our concern.

We will first describe a force control problem and then discuss mechanisms for sensing and controlling these forces. We will then define and discuss the concept of compliance and show two schema for developing compliant control.

## 11.1 A FORCE CONTROL PROBLEM

Figure 11.1 shows a typical force control problem, turning a crank. When programming a task such as this, one is initially tempted to use path control. After all, in Chapter 9, we learned how to make the hand smoothly pass through a path in space. Surely, we could generate a circular path and simply program the robot to follow that path. To see why force control might be preferred to position control in this situation, consider Figure 11.2, showing the desired position and true position of the crank handle. In that figure, due to errors in measurement, or eccentricity in the crank mechansim, the instantaneous planned

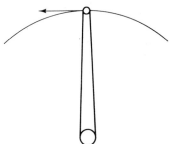

**Figure 11.1**   The problem of turning a crank is easily solved by applying a force in a tangential direction.

path is outside the true radius of the crank. If it is using position control, the robot will attempt to zero the positional error by stretching the crank handle.

This example shows one case of a basic problem with position control, that the controller cannot *comply* with the slightest variations from its predicted path.

Another example, Figure 11.3(a) shows this even more clearly. The task is defined (Brady, 1983) by

1. Place $B_3$ on $T$.
2. Move $B_3$ along $T$ until $B_3$ is against $B_2$'s left face.

A position control system would plan a path along the surface of $T$. However, if the surface of $T$ varied even slightly, the robot, moving

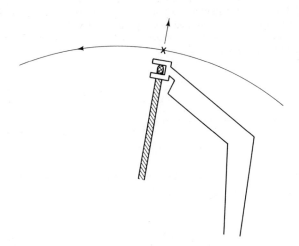

**Figure 11.2**   Due to measurement errors or eccentricity in the crank mechanism, the instantaneous planned path $(x)$ is outside the true radius of the crank handle. If using position control, the robot will attempt to zero this positional error by stretching the crank handle.

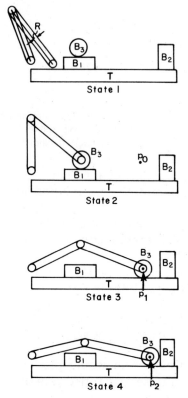

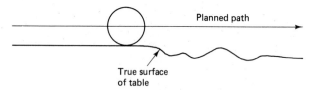

**Figure 11.3(a)**  (Brady) Representation of a scribing task for the two-link planar manipulator.  The robot is required to grasp the circular object $B_3$, representing a cutting stylus, and draw it along the surface $T$ until it is against the left face of $B_2$.  (*Courtesy MIT Press*)

**Figure 11.3(b)**  If the surface is not truly flat, a path control scheme will lose contact with the surface.

**Figure 11.3(c)**  If the surface tilts up, the path control scheme will attempt to press the part into the surface.

in a straight line, would either lose contact with the surface (Figure 11.3(b)) or would attempt to push $B_3$ into the surface (Figure 11.3(c)). A better approach to both these problems would be force control. We will return to the case of Figure 11.3 later, when we discuss hybrid control. For now, consider the crank turning problem:

To turn this crank effectively and be able to comply with inaccuracies in the model of the environment, we specify the task as

Exert a tangential force of magnitude $F_0$.

The force tangential to the rotation of the crank may be resolved into its Cartesian components as

$$f_x = f_0 \sin \theta \qquad (11.1a)$$

$$f_y = f_0 \cos \theta \qquad (11.1b)$$

Immediately, we observe that the force to be applied is a function of the position of the crank and, therefore, since the crank is turning, a function of time. The fact that the applied force is a function of arm configuration is true in general. Therefore, this applied force must be recomputed rapidly during a motion. This puts yet another real-time demand on our computational subsystem.

## 11.2 TRANSFORMING FORCES AND MOMENTS

In the previous section, we discussed the need for force control and showed two examples where it might be used. We have not addressed either the question of how to sense what the forces are, or how to apply forces. In this section, we will develop a relationship between the forces applied at the hand and the forces and torques applied by the actuators.

*Work* is force applied through a distance. We can derive a relationship between forces and torques through the principle of *virtual work*, that is, the work which would be done by applying a force **F** through an infinitesimal (vector) distance $\Delta \mathbf{x}$. In a robot, the virtual work done by the hand will be exactly equal to the virtual work done by the actuators applying forces (or torques) through infinitesimal distances (or angles). That is,

$$\mathbf{F}^\mathsf{T} \cdot \Delta \mathbf{X} = \boldsymbol{\tau}^\mathsf{T} \Delta \boldsymbol{\theta} + \mathbf{Q}^\mathsf{T} \Delta \boldsymbol{\theta} \qquad (11.2)$$

In this equation, **F** is the vector of Cartesian forces and moments; for example,

$$\mathbf{F} = \begin{bmatrix} f_x \\ f_y \\ f_z \end{bmatrix}$$

$\boldsymbol{\tau}$ is the vector of joint variables, either forces (for prismatic joints) or torques (for rotary joints), applied by the actuators; for example,

$$\boldsymbol{\tau} = \begin{bmatrix} f_r \\ T_{\theta_1} \\ T_{\theta_2} \end{bmatrix}$$

**Q** is the vector of joint forces or torques due to external forces such as gravity.

We know, from Chapter 8, that differential Cartesian motions are related to differential joint motions by the Jacobian matrix:

$$\Delta \mathbf{X} = J \Delta \boldsymbol{\theta} \tag{11.3}$$

Therefore, Eq. 11.2 becomes

$$\mathbf{F}^\mathsf{T} \cdot J \Delta \boldsymbol{\theta} = [\boldsymbol{\tau} + \mathbf{Q}]^\mathsf{T} \Delta \boldsymbol{\theta} \tag{11.4}$$

Since this is true for any $\Delta \boldsymbol{\theta}$, we have

$$\mathbf{F}^\mathsf{T} J = [\boldsymbol{\tau} + \mathbf{Q}]^\mathsf{T} \tag{11.5}$$

or

$$\mathbf{F}^\mathsf{T} = [\boldsymbol{\tau} + \mathbf{Q}]^\mathsf{T} J^{-1} \tag{11.6}$$

We thus relate the forces exerted in Cartesian space to the joint variables.

**Example 11.1   Force-torque Relations for the $\theta$-$r$ Manipulator**

The Jacobian of the $\theta$-$r$ manipulator is (Eq. 8.27)

$$J = \begin{bmatrix} \cos \theta & -r \sin \theta \\ \sin \theta & r \cos \theta \end{bmatrix} \tag{11.7}$$

The inverse can be expressed as Eq. 8.31 or as

$$J^{-1} = \begin{bmatrix} \cos \theta & \sin \theta \\ -\dfrac{1}{r} \sin \theta & \dfrac{1}{r} \cos \theta \end{bmatrix} \tag{11.8}$$

The vector **F** is the 2-vector of Cartesian forces

$$\mathbf{F} = \begin{bmatrix} f_x \\ f_y \end{bmatrix}$$

The joint forces are a force in the radial direction (since joint $r$ is prismatic), $f_r$, and a torque about $\theta$, $T_\theta$.

$$\boldsymbol{\tau} = \begin{bmatrix} f_r \\ T_\theta \end{bmatrix}$$

**Q** is the gravity loading.  The gravity loading in the radial direction is

$$Q_r = m_2 g \sin \theta \qquad (11.9)$$

and the gravity-induced torque about $\theta$ is

$$Q_\theta = (m_1 r_1 + m_2 r) g \cos \theta. \qquad (11.10)$$

Thus, Eq. 11.6, for this manipulator, becomes

$$\begin{bmatrix} f_x \\ f_y \end{bmatrix}^{\mathsf{T}} = \begin{bmatrix} f_r + m_2 g \sin \theta \\ T_\theta + (m_1 r_1 + m_2 r) g \cos \theta \end{bmatrix}^{\mathsf{T}} \begin{bmatrix} \cos \theta & \sin \theta \\ \dfrac{-1}{r} \sin \theta & \dfrac{1}{r} \cos \theta \end{bmatrix} \qquad (11.11)$$

Therefore,

$$f_x = (f_r + m_2 g \sin \theta) \cos \theta - \frac{T_\theta + (m_1 r_1 + m_2 r) g \cos \theta}{r} \sin \theta \qquad (11.12)$$

$$f_y = (f_r + m_2 g \sin \theta) \sin \theta + \frac{T_\theta + (m_1 r_1 + m_2 r) g \cos \theta}{r} \cos \theta \qquad (11.13)$$

To develop an intuitive understanding of the meaning of the equations derived in Example 11.1, let us evaluate them at

$$\theta = 0 \quad \text{and} \quad \theta = 90°$$

At $\theta = 0$,

$$f_x = f_r$$

$$f_y = \frac{T_\theta + (m_1 r_1 + m_2 r) g}{r} = \frac{T_\theta}{r} + \frac{(m_1 r_1 + m_2 r) g}{r}$$

At $\theta = 90°$,

$$f_x = \frac{-T_\theta}{r}$$

$$f_y = f_r + m_2 g$$

In Figure 11.4, these two conditions are shown:

At $\theta = 0$, the force in the $x$ direction is exactly the force exerted by the radial actuator.  The rotary actuator applies a torque $T_\theta$ which, acting through lever arm $r$, applies an upward force of $T_\theta/r$.  However, the actuator must also overcome the torque due to gravity applied to

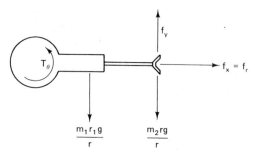

Figure 11.4(a)   $\theta$-$r$ manipulator at $\theta$ = 0.

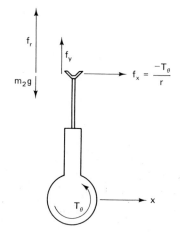

Figure 11.4(b)   $\theta$-$r$ manipulator at $\theta$ = 90°.

the two masses, leaving a resulting upward force at the hand of

$$f_y = \frac{T_\theta}{r} + \frac{(m_1 r_1 + m_2 r)g}{r}$$

(Note that the acceleration of gravity is defined positive in the downward direction).

At $\theta$ = 90°, the force in the $x$ direction is due entirely to the torque applied by the rotary actuator acting through the moment arm $r$. The radial actuator must overcome the gravity loading, leaving a resulting upward force of $f_r = m_2 g$. Note that in this position, mass $m_2$ has no effect.

Equations 11.12 and 11.13 provide us with a means for relating the outputs of the actuators with the forces applied by the hand, for the $\theta$-$r$ manipulator. Similar equations may be derived for full, six-degrees-of-freedom, manipulators. In the more general case, one must include not only forces, but also moments, applied by the hand. The techniques are a straightforward extension of the case just described. See Coiffet (1983) or Mason (1983) for more detail.

Given a desired force to be applied, the inverses of Eqs. 11.12 and 11.13 (see problem 1) tell us how to choose joint torques so that desired hand force will result. In an ideal, frictionless system, these would indeed be the resultant forces. Furthermore, in the presence of externally applied forces, these forces could be determined by sensing the motor torque. Thus, this application of the inverse kinematic transform can provide both sensing and control, in an ideal, rigid, reversible (back-drivable) system. Unfortunately, typical robots are neither rigid nor reversible. In many cases, external forces applied at the hand are almost unrelated to the motor torques; it, therefore, becomes necessary to sense the hand forces and moments directly.

## 11.3 DIRECT SENSING

Sensing of forces and moments, particularly in assembly applications, may be implemented by placing the workpiece on a sensor platform, rather akin to a very sophisticated scale. However, a more common strategy is to insert a sensor between the wrist and gripper.

There are various kinds of sensors, including peizoelectric, magnetic, and magnetostrictive devices and strain gauges (Binford, 1973). At the current state of the art, strain gauges have proven to be the most popular (Watson and Drake, 1975; Goto, Inoyama, and Takeyasu, 1974; Rosen and Nitzan, 1975; Groome, 1972; Bejczy, 1980; and Brunet and Hirzinger, 1983).

Figure 11.5 shows a typical wrist sensor (Coiffet, 1983). Measurement of the signals from the different sensors can be used to find the forces and moments using

$$
\begin{bmatrix} f_x \\ f_y \\ f_z \\ M_x \\ M_y \\ M_z \end{bmatrix} = \begin{bmatrix} 0 & 0 & k_{13} & 0 & 0 & 0 & k_{17} & 0 \\ k_{21} & 0 & 0 & 0 & 0 & k_{26} & 0 & 0 \\ 0 & k_{32} & 0 & k_{34} & 0 & k_{36} & 0 & k_{38} \\ 0 & 0 & 0 & k_{44} & 0 & 0 & 0 & k_{48} \\ 0 & k_{52} & 0 & 0 & 0 & k_{56} & 0 & 0 \\ k_{61} & 0 & k_{63} & 0 & k_{65} & 0 & k_{67} & 0 \end{bmatrix} \begin{bmatrix} w_1 \\ w_2 \\ w_3 \\ w_4 \\ w_5 \\ w_6 \\ w_7 \\ w_8 \end{bmatrix}
$$

$$(11.14)$$

where the $k_{ij}$ are constants characterizing gains of the sensors. (See also Shimano, 1978, for a more thorough discussion of the direct sensing and transform of forces.)

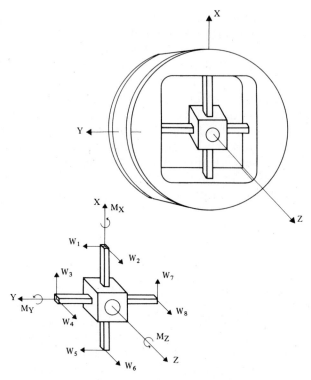

**Figure 11.5**   (Coiffet) Bejczy's sensitive wrist.

Some systems (Brunet and Hirzinger, 1983) include this transformation as part of the sensor interface package so that the user need
not concern himself or herself with such details. With such a sensor,
forces and moments may be measured at a point near the tool tip.
Fairly straightforward transformations then allow determination of true
forces at the tip.

## 11.4 COMPLIANCE

A *compliant robot system* is one that complies with externally generated forces, that is, one that modifies its motions in such a way as to
minimize some particular force or forces.  If the robot uses a force
sensor such as that described in the previous section, and modifies its
control strategy based on that sensor's output, the term *active compliance* is used to describe the behavior. On the other hand, if the robot's
gripper is constructed in such a way that the mechanical structure de-
forms to comply with those forces, the term *passive compliance* is used.

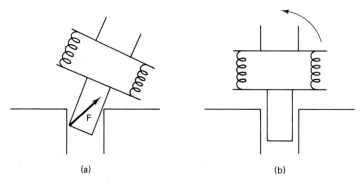

(a)                                              (b)

**Figure 11.6(a)**   Misalignment of the pin and hole will cause a force on the pin.
**(b)** If the robot moves in the correct direction, it can reduce the force on the pin.


### 11.4.1 Active Compliance

The general task of inserting a pin into a hole is representative of a large number of assembly operations such as pressing a bearing onto a shaft and has therefore been studied extensively. Figure 11.6 shows how a slight misalignment of the pin and the hole will result in a force on the pin and have both a force and a moment applied to the wrist sensor.

By moving in the correct direction (Figure 11.6(b)), the robot can reduce the force on the pin. Whitney (1983) has investigated the forces resulting from pin insertion in great detail and has included a careful analysis of the forces acting during insertion. For example, Figure 11.7 shows the forces acting on a pin being inserted into an unchamferred hole. The strategies of active compliance are closely related to those of hybrid control and will be discussed in more detail in that section.

### 11.4.2 Passive Compliance

As shown in Figure 11.8, another approach to compliance is to allow the hand to deform in such a way that the external forces are minimized. In that figure, springs are shown as providing the deformation.

The *remote center compliant* (RCC) gripper shown in Figure 11.9 provides an elegant solution to the peg-in-the-hole problem and all its variants. The supports indicated by the arrows in Figure 11.9 are flexible and bend to comply with the applied moments. Unlike springs, however, these supports undergo only lateral bending, not compression. Therefore, if there is no misalignment (i.e., the force applied is exactly along the shaft of the pin), there will be no bending, and full force will be applied to the pin. If a misalignment occurs, then some of the

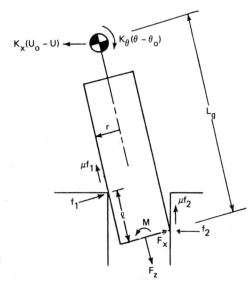

**Figure 11.7**  (Whitney 83) Forces acting
during two-point contact.

members will bend and rotate the tool tip to eliminate the bending
moment. Figure 11.10 describes a model of the RCC, except that, since
the RCC rotates about the tool tip, $L_g$ is nearly 0. Applications of the
RCC are analyzed in detail in Whitney (1983). Since the bending occurs
about the tool tip, which is remote from the gripper itself, the term
*remote center compliance* (Whitney, 1983) is used.


## 11.5 HYBRID CONTROL

In many applications, for example, those depicted in Figure 11.3, there
is a need for simultaneous control of position and force. Such control
is referred to as *hybrid control.* In general, a hybrid controller is one
that allows force to be commanded along certain Cartesian degrees of
freedom while allowing position to be commanded along the remaining
degrees of freedom (Mason, 1983).

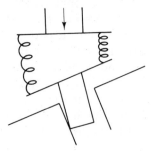

**Figure 11.8**  Passive compliance using a
spring-loaded wrist.

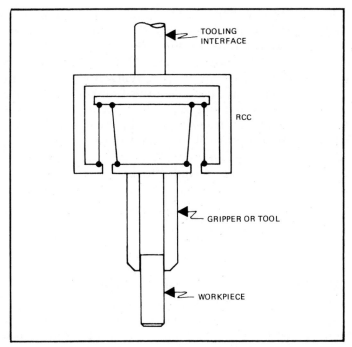

**Figure 11.9**    (Whitney 79) Remote-center compliant wrist. (*Courtesy of Robotics Today, used with permission.*)

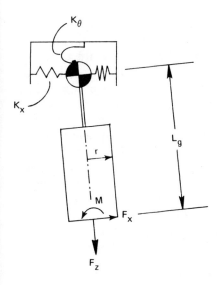

**Figure 11.10**    (Whitney 83) Rigid peg supported compliantly by lateral spring $K_x$ and angular spring $K$ at a distance $q$ from peg's tip. (*Courtesy ASME*)

The degrees of freedom that we are discussing are in a Cartesian coordinate system, but not necessarily the coordinate system of the manipulator hand or the force sensor. Rather, it is a Cartesian frame defined with respect to the task geometry. Takase, Inoue, and Hagiwara (1974) introduced the concept of specifying force constraints in a Cartesian frame. Paul (1938) has called this frame the *constraint frame* and has denoted it by the symbol [C].

In a hybrid controller, each actuator control signal is composed of several components—one for each force controlled degree of freedom in [C] and one for each position controlled degree of freedom. Raibert and Craig (1981) describe such a controller by

$$T_i = \sum_{j=1}^{N} \{\Gamma_{ij}[s_j \Delta f_j] + \Psi_{ij}[(1 - s_j)\Delta x_j]\} \tag{11.15}$$

where

$T_i$ = torque applied by the $i$th actuator
$\Delta f_j$ = force error in $j$th degree of freedom of [C]
$\Delta x_j$ = position error in $j$th degree of freedom of [C]
$\Gamma_{ij}$ and $\Psi_{ij}$ = force and position compensation functions, respectively, for the $j$th input and the $i$th output
$s_j$ = component of compliance selection vector

The compliance selection vector, **S**, is a binary $N$-tuple that specifies which degrees of freedom in [C] are under force control (indicated by $s_j = 1$) and which are under position control ($s_j = 0$). Figure 11.11 shows the resulting structure of a hybrid controller.

Two independent feedback loops provide the drive signals to the actuators. In Figure 11.11, the fact that all computations are relative to the constraint frame is made explicit by adding a superscript **C** on the left of each variable. The two feedback loops compute errors in both applied force and current position, with the bottom loop controlling force. The lower loop will be discussed in more detail.

The true force **F** is subtracted from the desired force $\mathbf{F}_d$, to produce a force error, $\mathbf{F}_e$. That error is a vector of six components, three forces and three moments. The block marked [$\mathscr{S}$] in Figure 11.11 serves to discard the components which are not to be force controlled. That is, if only the $y$ degree of freedom is to be controlled, all other elements of $\mathbf{F}_e$ are set to zero prior to being passed on to the coordinate transform block. This modified $\mathbf{F}_e$ is then converted to joint space using the appropriate coordinate transform. The resulting joint torques are then added to the torques resulting from the position control loop.

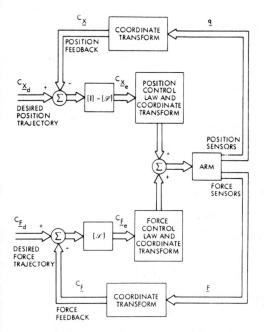

Figure 11.11   (Raibert) Conceptual organization of hybrid controller. (*Courtesy ASME*)

### Example 11.2  Hybrid Control of the $\theta$-$r$ Manipulator

Develop a hybrid control strategy for the $\theta$-$r$ manipulator. Assume that the constraint frame is coincident with the hand frame and that the force sensor provides its readings in the hand frame.

**Solution:**   The problem to be solved is shown in Figure 11.12; the robot is to move from the point $x = -a$ to $x = a$ while exerting a force $f_0$ in the $y$ direction. No other constraints are defined. In particular, no control over velocity will be required. In this example problem, we will also assume that the manipulator is moving in a horizontal plane so that gravity may be ignored.

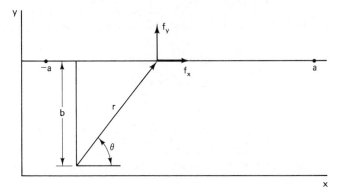

Figure 11.12   Hybrid control of the $\theta$-$r$ manipulator.

We define the joints and forces as follows:

$$\tau_1 = f_r = \text{force applied by the radial actuator}$$
$$\tau_2 = T = \text{torque about } \theta$$
$$f_1 = f_x = \text{force in the } x \text{ direction}$$
$$f_2 = f_y = \text{force in the } y \text{ direction}$$

In this case, the compliance selection vector $\mathbf{S}$ defined in Eq. 11.15 is

$$\mathbf{S} = [0, 1]^\mathsf{T}$$

since the second degree of freedom, $y$, is defined to be force controlled.
First, we find the position and force errors.

$$\Delta \mathbf{X}(t) = \mathbf{X}_d - \Lambda(\boldsymbol{\theta}(t))$$
$$\Delta \mathbf{F}(t) = \mathbf{F}_a - \mathbf{F}(t)$$

where $\mathbf{X}_d$ is the vector of desired Cartesian positions

$$\mathbf{X}_d = \begin{bmatrix} x_d \\ y_d \end{bmatrix} \quad \text{or in this example} \quad \begin{bmatrix} a \\ b \end{bmatrix}$$

$\mathbf{X}(t) = \Lambda(\boldsymbol{\theta}(t))$ is the vector of current Cartesian positions. $\boldsymbol{\theta}(t)$ is the vector of joint variables

$$\boldsymbol{\theta}(t) = \begin{bmatrix} r(t) \\ \theta(t) \end{bmatrix}$$

$\Lambda$ is the vector form of the forward kinematic transform operator. The individual elements of $\Lambda$ are the $\Psi_{ij}$ given in Eq. 11.15.

$\mathbf{F}_d$ is the desired force vector

$$\mathbf{F}_d = \begin{bmatrix} f_{dx} \\ f_{dy} \end{bmatrix} \quad \text{or in this example} \quad \begin{bmatrix} 0 \\ f_0 \end{bmatrix}$$

$\mathbf{F}$ is the true force measured by the sensor

$$\mathbf{F} = \begin{bmatrix} f_x \\ f_y \end{bmatrix}$$

Applying these definitions, we find expressions for the components of position and force error

$$\Delta x = a - r \cos \theta$$
$$\Delta y = b - r \sin \theta$$
$$\Delta f_x = 0 - f_x$$
$$\Delta f_y = f_0 - f_y$$

Next, we apply the vector $\mathbf{S}$ as a mask to separate the force controlled and position controlled degrees of freedom and derive the Cartesian error signals, which we define as follows:

$$\mathbf{X}_e = \begin{bmatrix} x_e \\ y_e \end{bmatrix}, \text{ the error in position}$$

$$\mathbf{F}_e = \begin{bmatrix} f_{xe} \\ f_{ye} \end{bmatrix}, \text{ the error in force}$$

$$\boldsymbol{\theta}_e = \begin{bmatrix} r_e \\ \theta_e \end{bmatrix}, \text{ the error in joint position}$$

$$\boldsymbol{\tau}_e = \begin{bmatrix} f_{re} \\ T_{\theta e} \end{bmatrix}, \text{ the error in force or torque}$$

$$x_e = (1 - s_1)\Delta x = a - r\cos\theta$$

$$y_e = (1 - s_2)\Delta y = 0$$

$$f_{xe} = s_1 \Delta f_x = 0$$

$$f_{ye} = s_2 \Delta f_y = f_0 - f_y$$

Transforming from Cartesian to joint space, we find the errors in joint angle due to $x_e$ and $y_e$. We do this most simply by observing that $\mathbf{X}_e$ is a differential change, and using the Jacobian to find the corresponding (approximate) differential change in $\boldsymbol{\theta}$.

$$\boldsymbol{\theta}_e = J^{-1}\mathbf{X}_e$$

or

$$\boldsymbol{\theta}_e = \begin{bmatrix} r_e \\ \theta_e \end{bmatrix} = \begin{bmatrix} x/r & y/r \\ -y/r^2 & x/r^2 \end{bmatrix} \begin{bmatrix} a - r\cos\theta \\ 0 \end{bmatrix}$$

so

$$r_e = (x/r)(a - r\cos\theta)$$

$$\theta_e = (-y/r^2)(a - r\cos\theta)$$

Similarly, we use the Jacobian to relate forces

$$\boldsymbol{\tau}_e = J^T \mathbf{F}_e$$

$$= \begin{bmatrix} f_{re} \\ T_{\theta e} \end{bmatrix} = \begin{bmatrix} \cos\theta & \sin\theta \\ -r\sin\theta & r\cos\theta \end{bmatrix} \begin{bmatrix} 0 \\ f_0 - f_y \end{bmatrix}$$

so

$$f_{re} = (f_0 - f_y)\sin\theta$$

$$T_{\theta e} = (f_0 - f_y) r\cos\theta$$

We now have four error signals, all in joint space. We may use any control strategy we wish to move the joints to reduce that error.[*]

For simplicity, we will use PE (proportional error) on force and PD (proportional derivative) on position. By choosing these control strategies, we have defined

---

[*]Raibert and Craig (1981) have performed realistic experiments using hybrid control on an electric arm. They found it necessary to use more complex controllers (proportional derivative and proportional integral derivative) to get satisfactory performance. Since our model ignores friction and gravity, we use an unrealistically simple control.

four control signals:

$$u_{rp} = k_{rp}(r_e) + k_{rpd}(\dot{r}_e)$$

$$u_{\theta p} = k_{\theta p}(\theta_e) + k_{\theta pd}(\dot{\theta}_e)$$

$$u_{rf} = k_{rf}(f_{re})$$

$$u_{\theta f} = k_{\theta f}(T_{\theta e})$$

where the gains are the usual servo feedback gains.

Finally, the drive to be applied to the radial actuator is

$$u_r = u_{rp} + u_{rf}$$

and

$$u_\theta = u_{\theta p} + u_{\theta f}$$

## 11.6 CONCLUSION

In this chapter, we have discussed compliant control and force or hybrid control of robots. We have seen that compliance can be gained either actively by a servo process or passively by a compliant tool.

Hybrid controllers can be used to implement compliant control functions. We have described the general structure of hybrid controllers and have gone through one example in detail. A significant computational problem exists with implementation of hybrid control strategies due to the requirement for continuous coordinate transforms. Fortunately, when performing hybrid control functions, the robot is generally moving slowly, allowing a reasonable update rate from the computer.

## 11.7 SYNOPSIS

### Vocabulary

You should know the definition and application of the following terms:

active compliance
compliance
constraint frame
direct sensing
force control
hybrid control
passive compliance

remote center compliance
virtual work

**Notation**

| Symbol | Meaning |
|--------|---------|
| **F** | A force (3-vector) |
| $\Delta \mathbf{X}$ | An infinitesimal distance |
| $\boldsymbol{\tau}_i$ | A torque (3-vector) about the $i$ actuator |
| $\Delta \boldsymbol{\theta}$ | An infinitesimal rotation |
| Q | Vector of torque due to external forces |
| $f_i$ | Force in the $i$ direction |
| $J$ | Jacobian matrix |
| $g$ | Acceleration of gravity |
| $T_\theta$ | Torque about the $\theta$ axis |
| $\Delta f_j$ | Force error in the $j$th direction |
| $\Delta x_j$ | Position error in the $j$th direction |
| $\Gamma_{ij}$ | Force compensation function |
| $\Psi_{ij}$ | Position compensation function |
| **S** | Compliance selection vector |
| $\Lambda$ | An operator, the vector form of the forward kinematic transform |

## 11.8 REFERENCES

BEJCZY, K.  Smart sensors for smart hands. *Progress in Astronautics and Aeronautics,* American Institute of Aeronautics and Astronautics, pp. 275–304, 1979.

BINFORD, T. D.  Sensor system for manipulation. *Proceedings of the First Conference on Remotely Manned Systems,* 1973.

BRADY, M.  Basics of robot motion, planning and control.  In *Robot motion, planning, and control,* pp. 1–50, Brady M., Hollerbach J., Johnson T., Lozano-Pérez T., and Mason M., eds., Cambridge, Mass.: M.I.T. Press, 1983.

BRUNET, U. and HIRZINGER, G.  Fast and self-improving compliance using digital force-torque-control.  4th International Conference on Assembly Automation, Tokyo, October, 1983.

COIFFET, P.  *Robot technology.*  Vol. 2: *Interaction with the environment.*  Englewood Cliffs, N.J.: Prentice-Hall, 1983.

GOTO, T., INOYAMA, T., and TAKEYASU, K.  Precise insert operation by tactile controlled robots. *Proceedings of 4th International Symposium on Industrial Robots,* Tokyo, 1974.

GROOME, R. C.   Force feedback steering of a teleoperator system, Technical Report 575. Draper Labs, Cambridge, Mass. 1972.

MASON, M. T. Compliant motion.   In *Robot motion, planning, and control*, pp. 305–322.   Brady, Hollerbach, Johnson, Lozano-Pérez, and Mason, eds.   Cambridge, Mass.: M.I.T. Press, 1983.

PAUL R. P. *Robot manipulators.* Cambridge, Mass.: M.I.T. Press, 1981.

RAIBERT, M. C. and CRAIG, J. J.   Hybrid position/force control of manipulators. *Journal of Dynamic Systems, Measurement, and Control,* June 1981; also in *Robot motion, planning, and control*, pp. 419–438.   Brady, Hollerbach, Johnson, Lozano-Pérez, and Mason, eds. Cambridge, Mass.: M.I.T. Press, 1983.

ROSEN, C. A. and NITZAN, D.   Developments in programmable automation. *Manufacturing Engineering,* September 1975.

SHIMANO, B.   The kinematic design and force control of computer-controlled manipulators. AI Lab., Stanford University, Memo 313, March 1978.

TAKASE, K. H., INOUE, K., and HAGIWARA, S.   The design of an articulated manipulator with torque control ability. *Proceedings of the 4th International Symp. on Industrial Robots,* Tokyo 1974.

WATSON, P. C. and DRAKE, S. H.   Pedistal and wrist force sensors for automatic assembly. *Proceedings of the 5th International Symposium on Industrial Robots,* IIT, Chicago, September 1975.

WHITNEY, L. E.   Quasi-static assembly of compliantly supported rigid parts. *Journal of Dynamic Systems, Measurement, and Control,* pp. 65–77, March 1982; also in *Robot motion, planning and control*, pp. 439–471.   Brady, Hollerbach, Johnson, Lozano-Pérez, and Mason, eds. M.I.T. Press, 1983.

## 11.9 PROBLEMS

1. Equation 11.11 gives the force exerted by the hand of the $\theta$-$r$ manipulator in terms of the forces and torques applied by the joint actuators. Invert this solution. That is, find an expression that gives the joint forces and torques as a function of the Cartesian forces on the hand.

2. Figure 11.12 illustrates a hybrid control application. The solution given in Example 11.2 ignores friction and gravity. Assume that the manipulator has mass $m$, acting in the negative $y$ direction (as defined by Figure 11.12), and re-derive the results of Example 11.2.

# 12

---

# *SENSORS*

Chapters 11, 12, and 13 of this book are concerned with sensors and their use. In Chapter 11, we discussed force control and determined that precise control of force required a good quality force-torque sensor, either mounted on the wrist of the robot or under the workpiece. In Chapter 13, we will address computer vision, the principal noncontact sense in robotics.

In this chapter, we will describe several other sensors that may be used in robot systems. These sensors all exhibit "local" properties, in that they sense the properties of a surface which is touching or nearly touching the robot's hand. Use of such sensors allows the robot to interact with its environment in an adaptive, "intelligent" way.

This chapter differs from most of the remainder of this book because as yet there exists neither paradigms for the design of such sensors nor well-structured techniques for choosing the appropriate sensor for particular applications. In the absence of paradigms, we will simply describe several sensors and discuss typical properties and potential applications.

## 12.1 TOUCH SENSORS*

The need for touch sensors occurs in many robotics applications, from picking oranges to loading machines (see Harmon, 1980 for a good

*The author is grateful to Jack Rebman of the Lord Corporation for his assistance in the preparation of this chapter.

summary and review). Probably the most important application currently is the general problem of locating, identifying, and organizing parts that need to be assembled. This application frequently employs computer vision systems such as those described in Chapter 13. However, many cases require positional information that cannot be provided by computer vision because of two major deficiencies inherent in computer vision systems.

The first deficiency is accuracy. In a typical parts handling operation (Page, Snyder, and Rajala, 1983), the vision system could position the robot to within $\frac{1}{4}$ inch while maintaining a field of view of 5 feet horizontally. Such positioning accuracy is quite good for a vision system, but it may not be adequate for the precise positioning needed to insert a part into a machine tool.

The second major deficiency of computer vision arises from the fairly obvious fact that a computer vision system cannot see behind the part. If gripping relies on somehow reaching around the part, blind gripping may lead to damage of the parts to the rear.

For both these reasons, a "smart" gripper is often needed, even in systems with vision. Such sensory capabilities may take on a variety of forms, as will be demonstrated in this section. Another survey of hand-mounted sensors may be found in Bejczy (1977).

The more sophisticated electronic sensors emulate the human sense of touch. The physiological sense of touch has two distinct aspects: the *cutaneous sense*, which refers to the ability to perceive textural patterns encountered by the skin surface; and the *kinesthetic sense*, which refers to the ability to determine forces and moments.

A touch sensor system thus includes the capability to detect such things as (Rebman and Trull, 1978)

1. Presence
2. Part shape, location, orientation
3. Contact area pressure and pressure distribution
4. Force magnitude, location, and direction
5. Moment magnitude, plane, and direction

The major components of a tactile sensor system are

1. A touch surface
2. A transduction medium, which converts local forces or moments into electrical signals.
3. Structure
4. Control/interface

Hill and Sevard (1973) describe a gripper in which each finger is equipped with seven sensitive panels to be used in collision detection.

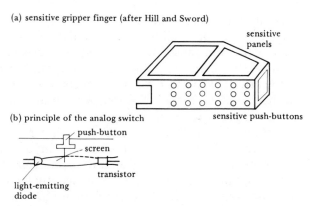

(a) sensitive gripper finger (after Hill and Sword)

(b) principle of the analog switch

**Figure 12.1(a)**    (Coiffet) Sensitive finger.

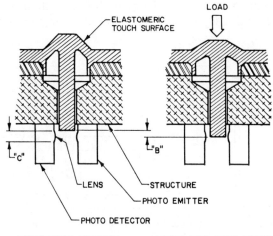

ZERO DEFLECTION        MODERATE DEFLECTION

**Figure 12.1(b)**    (Rebman & Trull) Sensitive site detail. (*Courtesy ASME*)

In addition, the inside of each finger is covered with an array of 18 pushbottons. As each button is pressed, it partially interrupts a beam of light (Figure 12.1). The more pressure on the button, the more the beam is blocked. This signal is detected and provides a rough outline of the object being grasped. A refined and robust sensor (Figure 12.1(b)) using this concept has been developed for industrial applications by Lord Corporation (Rebman, 1983).

The concept of sensitive panels has been extensively used by (Kinoshita and Mori, 1972), who describes a multifingered, articulated, hand with sensitive panels on the inside of each link of each finger. As might be expected, the transformations to relate sensor output to three-dimensional surface contour are very complex.

Sensors described by Peruchon (1979) and Page, Puch, and Heginbotham (1976) use arrays of small metal rods and detect the deflection
of each rod. The Peruchon sensor uses the rod as a variable core in an
inductor and detects the change in inductance as the rod is brought into
contact with the surface. The Page sensor detects not the height, but
the variation in height of each rod. The sensor is brought down slowly
on the part while the stand is vibrated (Figure 12.2). When all the rods
move together, the part is in full contact with the gripper.

Several researchers (Bejczy, 1980; St. Clair and Snyder, 1978) have
reported on experiments with conductive elastomers or *artificial skins.*
While they vary in details, these systems are conceptually similar, and
we describe only the system by St. Clair and Snyder (1978).

A conductive elastomer is a rubberlike material whose electrical
conductivity changes (locally) when compressed. In this system, the
elastomer was laid over a printed circuit board etched with 16 pairs of
concentric rings. Each pair of rings formed a sensing element (Figure

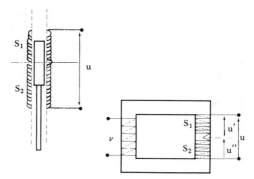

**Figure 12.2(a)**   (Coiffet) Principle of the Peruchon sensor.

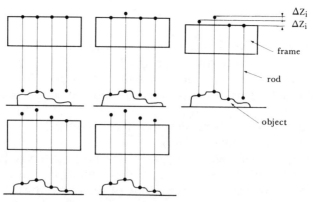

**Figure 12.2(b)**   (Coiffet) The Page sensor: for every increment of frame movement the rods that have undergone movement $\Delta Z_i$ are counted.

12.3).  The outer rings were wired in parallel four at a time to form four rows.  The center rings were likewise wired to form four columns. Thus, any sensor could be $x$-$y$ addressed.

A current was allowed to flow from the inner to outer selected rings, through the elastomers.  By measuring the current-voltage ratio, the local conductivity of the elastomer could be determined, and hence the applied force.

The authors then applied pattern recognition techniques to determine edges, points, and other features.

The major problem reported by these authors, as well as others who have experimented with these materials, is the repeatability of the measurement.  Once compressed, the material takes a very long time to recover its original characteristics.

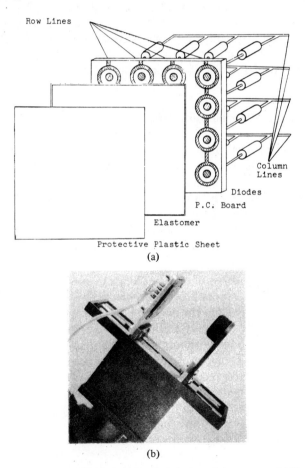

(a)

(b)

**Figure 12.3**    (St. Clair & Snyder) (a) Exploded view of the sensor.  (b) The tactile sensor.  (*Courtesy 1978 IEEE*)

## 12.2 PROXIMITY SENSORS

In this section, we will discuss devices that allow the robot to determine some properties of a surface without actually contacting that surface. The property most often detected is simply the presence or absence of a surface. This function provides collision avoidance. In addition, proximity sensors provide information about approach conditions, so that appropriate deceleration and maneuvering may be performed prior to grasping.

### 12.2.1 Optical Proximity Detectors

An optical proximity detector such as the one depicted in Figure 12.4 can provide a simple binary signal, indicating the presence of an object. Calibration is required for each object, since the distance at which the sensor's output goes true will depend on the reflectivity of the surface. Such a sensor can be used either for collision avoidance or to signal the robot when it has reached a precise distance from the object.

### 12.2.2 Optical Ranging Using Reflectance

By projecting a calibrated light source onto the object and measuring the reflected intensity, one could conceivably determine the distance from the sensor to the object. Successful use of such a sensor requires some understanding of the ways in which light is reflected from a surface.

The most familiar form of reflection, purely specular reflection, is shown in Figure 12.5(a). In specular reflection, the emitted light ray forms an angle with the surface normal exactly equal to the incident ray.

If the incident and emitted angles are not identical, then some *scattering* is said to have occurred. The most often discussed form of scattering, *Lambertian scattering*, is depicted in Figure 12.5(b). In a surface which exhibits Lambertian scattering, the net effect of all the

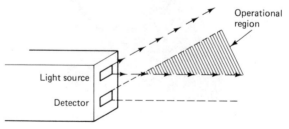

**Figure 12.4** Optical proximity detector containing a light source (typically an LED), and a light detector.

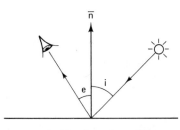

Figure 12.5(a)   Specular reflection.

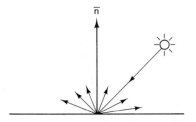

Figure 12.5(b)   Lambertian scattering.

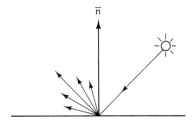

Figure 12.5(c)   Forward scatter.

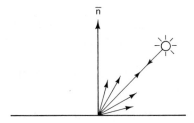

Figure 12.5(d)   Backscatter.

processes occurring is that the brightness observed is independent of viewing angle and depends only on the angle between the surface normal and the incident ray.   That is, $E = a_0 I \cos i$, where $E$ is the observed brightness, $I$ is the incident light intensity, and $i$ is the angle between the incident ray and the surface normal.   This fact is utilized in some computer vision work to determine surface orientation.   See Horn (1975) or Ray (1981) for details.

Other types of scattering are shown in Figures 12.5(c) and (d).   In fact, most surfaces exhibit a mixture of these reflectivity properties. This fact, coupled with other difficulties in modeling, makes it more effective to calibrate a proximity sensor than to attempt to develop a good analytic model.   Figure 12.6 shows such a sensor, with both light

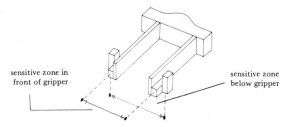

Figure 12.6   (Coiffet) Example of sensors attached to a manipulator gripper.

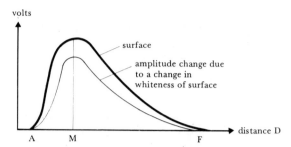

**Figure 12.7**    (Coiffet) Typical response of an infrared detector as a function of distance from a surface.

source and detector mounted on the gripper. Figure 12.7 shows a typical response characteristic of such a sensor. This curve will be slightly distorted if the surface is not plane. Coiffet (1983) points out three difficulties in using such a sensor:

1. The whiteness of the surface (by trial and error) has to be known. This is usually the case in industrial applications.
2. Except for the singularity at the top of the curve, the same signal can denote two possible distances.
3. The axis of the sensor must be normal to the surface.

. . . these proximity sensors are mostly used for the detection of the presence of an object in the volume, scanned only in rare cases to measure distance, and exceptionally, in recognition.

Bejczy (1980) however, reports on an improved sensor that eliminates the double-valued reading and provides improved performance by using fiber optics and more sophisticated signal processing. That sensor provides an effective sensing range of 7 to 8 cm.

### 12.2.3 Triangulation Proximity Sensors

If a plane of light is projected on a scene at a known angle and is observed from a known position, the light source, sensor, and object form a *range triangle* as shown in Figure 12.8(a). The projection angle and light source–sensor distance are known, and the observation angle is measured. From this, the distance, $d$, to the object may be determined.

This technique, known as *light striping*, is very popular in industrial machine vision. The same technique may be scaled down as shown in Figure 12.8(b) to a sensor mounted on the hand, utilizing a CCD sensor and a single light stripe.

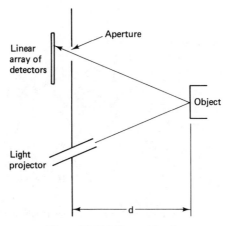

**Figure 12.8(a)** Range triangle.

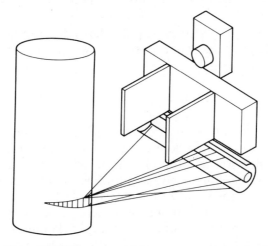

**Figure 12.8(b)** Camera and light source mounted on hand.

## 12.3 ULTRASONIC RANGING

An ultrasonic transducer emits a pulse of high-frequency sound and then listens for the echo.  Since sound in air travels at slightly under 1 foot per millisecond, the elaspsed time between initial transmission and echo detection can then be converted to distance.

The material or the topography could cause a pulse at any particular frequency to be canceled and, therefore, not echoed.  For this reason, most transducers actually transmit a "chirp" consisting of a range of frequencies.  For example, the Polaroid electrostatic transducer transmits a chirp consisting of four ultrasonic frequencies, 60 kHz (kilohertz), 57 kHz, 53 kHz, and 50 kHz.

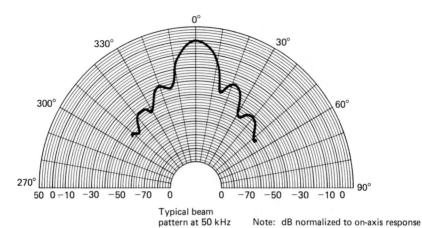

Typical beam
pattern at 50 kHz          Note:  dB normalized to on-axis response

**Figure 12.9**  Typical beam pattern at 50 kHz.  Note: db normalized to on-axis response.  Note: curves are representative only.  Individual responses may differ. (*Courtesy Polaroid*)

Attenuation of an ultrasonic pulse depends on frequency and path length.  Specifically, the sound pressure level at a distance $D$ from the sensor is

$$p = p_0 \, \frac{1}{D} \, e^{-\alpha D} \, .$$

where $\alpha$ is the absorption coefficient in air.  $\alpha$ increases with increasing frequency.  At 50 kHz, for example, the signal power returned at 3 feet is a million times stronger than is the signal returned at 35 feet.  For this reason a variable gain receive amplifier is a requirement.

Figure 12.9 shows the beam pattern of the Polaroid transducer. In general, any object subtending an angle of 2–4° with respect to the sensor will cause a return.  More specifically, Polaroid* specifies the performance test as follows:

At 4 feet, a sphere 2 inches in diameter will be detected within an acceptance angle of 8–14°, and at 15 feet, a sphere 12 inches in diameter will be detected within an acceptance angle of 8–19°.

The accuracy of this transducer is better than 1 percent of the range.  That is, an object 3 feet away can consistently be located to within $\frac{1}{2}$ inch.

Such sensors are most appropriate in robotic applications for detecting potential collisions and avoiding them and for approximate ranging, in the absence of vision.  For example, an ultrasonic sensor

---

*The author is grateful to Mr. Olin Brown of Polaroid for his assistance in describing the ultrasonic range sensor.

could get the gripper fairly close to the object, and then an optical proximity detector mounted on the hand could be used to provide fine ranging for precise positioning.

## 12.4 CONCLUSION

In this chapter, we have described several sensors and their potential applications in robotics. Ultrasonic sensors provide a look at the object from a distance of several feet and allow the robot to position the hand relatively close to the part, close enough for devices such as optical proximity detectors to become active. Such detectors can be used to position the hand until it is almost contact with the part. Finally, touch sensors, either pressure transducers, arrays of switches, or a mix of the two, can identify the part surface contours.

Both ultrasonic and optical proximity detectors can make serious errors when confronted with parts that have unusual shapes. For example, the part in Figure 12.10 has a small rod extending above the surface. That rod may be too small to be detected by an ultrasonic sensor and may escape the narrow field of view of a proximity detector. Such problems may be handled on an ad hoc basis by, for example, sweeping the hand in a spiral path searching for the rod or, in some cases, by computer vision, the topic of the next chapter.

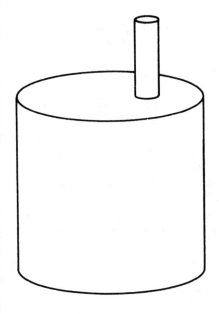

Figure 12.10

## 12.5 SYNOPSIS

**Vocabulary**

You should know the definition and application of the following terms:

artificial skin
cutaneous
elastomer
kinesthetic
Lambertian
light striping
optical proximity detector
reflectance ranging
scattering
specular
transduction
triangulation
ultrasonic

## 12.6 REFERENCES

BEJCZY, A. K. Effect of hand-based sensors on manipulator control performance. In *Mechanism and Machine Theory*, Vol. 12, pp. 547–567. New York. Pergamon Press, 1977.

———. Smart sensors for smart hands. In special issue of *Progress in Astronautics and Aeronautics* on *Remote Sensing of Earth from Space–Role of Smart Sensors*, pp. 275–304, 1980.

COIFFET, P. *Robot Technology*, Vol. 2, Englewood Cliffs, N.J.: Prentice-Hall, 1983.

HARMON, L. D. "Touch Sensing Technology: A Review," Tech. Report No. MSR80-03 pp. 57, Society of Manufacturing Engineers, 1980.

HILL, J. W., and SWORD, A. J. Manipulation based on sensor-directed control: An integrated end effector and touch sensing system, *Proceedings of the 17th Annual Human Factor Society Convention*, Washington, D.C., Oct. 1973.

HORN, B. Obtaining shape from shading information. In *The Psychology of Computer Vision*, pp. 115–156. P. Winston, ed. New York: McGraw-Hill, 1975.

KINOSHITA, G., and MORI, M. Design method for spacing the receptor elements on an artificial tactile sense with multielements. *Transactions of the Society of Instrument and Control Engineers*, 8(5), 1972.

PAGE, S., PUCH, A., and HEGINBOTHAM, W. Novel techniques for tactile sensing in a three-dimensional environment. *Proceedings of the 3rd Conference on Industrial Robot Technology*, University of Nottingham, 1976.

PAGE, N., SNYDER, W., and RAJALA, A. Turbine blade image processing system. First International Conference on Software for Robots, Liege, Belgium, May, 1983.

PERUCHON, E. "Contribution to Carrying Out Artificial Tactile Functions Based on Analysis of the Human System." Thesis, Montpellier, France, 1979.

RAY, R. Error analysis of surface normals determined by radiometry. In *General methods to enable robots with vision to acquire, orient, and transport workpieces.* Report to the National Science Foundation, University of Rhode Island, December 1981.

REBMAN, J., and TRULL, N. A robot tactile sensor for robot applications. ASME Conference on Computers in Engineering, Chicago, August, 1983.

ST. CLAIR, J., and SNYDER, W. E. Conductive elastomers as a sensor for industrial parts handling equipment. *IEEE Transactions on Instrumentation and Measurement*, IM27, March, 1978, pp. 94–99.

## 12.7 PROBLEMS

1. Analyze carefully using a robot to pick oranges. Under what conditions could vision help? What type of tactile sensor would seem appropriate and on what kind of gripper?

2. Propose an application of robotics that could make use of a sensor that provides only information about the cutaneous sense.

3. Propose an application of robotics that could make use of a sensor that provides only kinesthetic information.

4. Analyze the feasibility of the sensor shown in Figure 12.1(b). What are its limitations? Compare that sensor with the sensor of Figure 12.3. Which do you think is more robust? Why? (You might consider such properties as hysteresis, work hardening, reliability, and maintainability.)

5. Consider the problem of using an optical proximity detector to detect the presence of metallic parts. Describe the properties of the metallic surface which would allow such a device to be effective or not effective.

6. In Section 12.3, the claim is made that "At 50 kHz, the signal power returned at 3 feet is a million times stronger than the signal returned at 35 feet." Verify this.

# 13

---

# *COMPUTER VISION*

In this chapter, we will introduce the basic concepts in the field of study known as *computer vision*. This discipline emphasizes the development of techniques which allow a computer to recognize or otherwise understand the content of a picture. Numerous books have been written on this subject, and we could not hope to do it justice in one chapter. Therefore, we concentrate on a subdiscipline called IMV (*industrial machine vision*), which includes robot vision.

Researchers in the field of IMV concentrate their efforts on problems appropriate to the industrial environment. In such an environment, one may be able to control the background, the lighting, the camera position, or other parameters. Such control may allow the use of techniques that would be inappropriate to a general-purpose vision system.

In this chapter, the reader will be introduced to the concept of an electronic image and its digital representation. Then, several strategies for processing such images will be covered. These strategies will allow the computer to make use of image information to guide a robot.

## 13.1 FUNDAMENTALS

The first section in this chapter will show the reader how digital images may be acquired and represented in the memory of a computer. The operations which the computer must perform to make use of that information for industrial applications will be presented in later sections.

### 13.1.1  The Formation of a Digital Image

The imaging literature is filled with a variety of imaging devices, including dissectors, flying spot scanners, videcons, orthicons, plumbicons, CCDs (charge-coupled devices), and others (Castleman, 1977; Chien and Snyder, 1975). These devices differ both in the ways in which they form images and in the properties of the images so formed. However, all these devices convert light energy to voltage in similar ways. Since our intent in this chapter is to introduce the reader to the fundamental concepts of image analysis, we will choose one device, the videcon, and discuss the way in which digital images are formed using such a device.

#### Image Formation With a Silicon Videcon

As shown in Figure 13.1, a lens is used to form an image on the faceplate of the videcon. When a photon of the appropriate wavelength strikes the special material of the faceplate, a quantum of charge is created (an electron-hole pair). Since the conductivity of the material is quite low, these charges tend to remain in the same general area where they were created. Thus, to a good approximation, the charge, $q$, in a local area of the faceplate follows

$$q = \int_0^{t_f} i\,dt \qquad (13.1)$$

where $i$ is the incident light intensity, measured in photons per second. If the incident light is a constant over the integration time, then $q = it_f$, where $t_f$ is called the frame time.

The mechanism for reading out the charge, be it electron beam, as in the videcon, or charge coupling, as in CCD cameras, is always designed so that as much of the charge is set to zero as possible. We start the integration process with zero accumulated charge, build up the charge

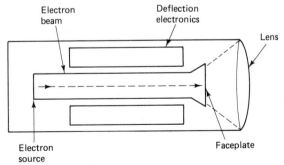

**Figure 13.1**   Videcon camera.

at a rate proportional to local light intensity, and then read it out. Thus, the signal read out at a point will be proportional to both the light intensity at that point and to the amount of time between read operations.

Since we are interested only in the intensities and not in the integration time, we remove the effect of integration time by making it the same everywhere in the picture. This process, called *scanning*, requires that each point on the faceplate be interrogated and its charge accumulation zeroed, repetitively and cyclically. Probably the most straightforward, and certainly the most common way in which to accomplish this is in a top-to-bottom, left-to-right scanning process called *raster scanning* (Figure 13.2).

In the videcon, an electron beam is used to neutralize the accumulated (positive) charge. Cancellation of the charge causes a current to flow in the circuit proportional to the charge neutralized. Deflection electronics steers the electron beam across the faceplate in a horizontal line. The beam is then shut off and repositioned at the left-hand end of the next, lower, line. The time the beam is off is referred to as *blanking*. This process is repeated until the entire faceplate has been swept clean of charge. However, while the beam is busy neutralizing charge at the bottom of the faceplate, charge is once again building up at the top. Since charge continues to accumulate over the entire surface of the videcon faceplate at all times, it is necessary for the beam to return immediately to the top of the faceplate and begin scanning again. The scanning process is repeated many times each second. In American television, the entire faceplate is scanned once every 33.33 ms (milliseconds) (in Europe, the frame time is 40 ms).

To compute exactly how fast the electron beam is moving, we compute

$$\frac{1\ \text{sec}}{30\ \text{frame}} \div \frac{525\ \text{lines}}{\text{frame}} = 63.5\ \mu\text{s/line} \qquad (13.2)$$

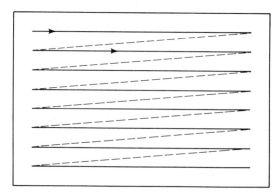

**Figure 13.2**   Raster scanning:  Active video is indicated by a solid line, blanking (retrace) by a dashed line. This simplified figure represents noninterlaced scanning.

Sixty-three and a half microseconds ($\mu$s) to scan one line of the picture is fairly fast. (Using the European standard of 625 lines and 25 frames per second, we arrive at almost exactly the same answer, 64 $\mu$s per line.) This 63.5 $\mu$s includes not only the active video signal but also the blanking period, approximately 18 percent of the line time. Subtracting this *dead time*, we arrive at the active video time, 52 $\mu$s per line.

Figure 13.3 shows the output of a television camera as it scans three successive lines. One immediately observes that the raster scanning process effectively converts a picture from a two-dimensional signal to a one-dimensional signal, where voltage is a function of time. Figure 13.3 shows both *composite* and *noncomposite video signals*, that is, whether the signal does or does not include the sync and blanking timing pulses.

The sync signal, while critical to operation of conventional television, is not particularly relevant to our understanding of digital image processing at this time. The blanking signal, however, is the single most important timing signal in a raster scan system. *Blanking* refers to the time that the electron beam is shut off. There are two distinct blanking events: *horizontal blanking*, when the beam moves from the end of one line to the start of the next, and *vertical blanking*, when the beam moves from the bottom of the picture to the top in preparation for a new scan. In a digital system, both blanking events may be represented by pulses on separate digital wires. Composite video is constructed by shifting these special timing pulses negative and adding them to the video signal.

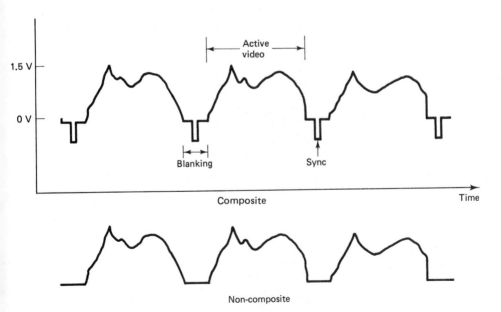

**Figure 13.3**    The video signal.

Now that we recognize that horizontal blanking signifies the beginning of a new line of video data, we can concentrate on that line and learn how a computer might acquire the brightness information encoded in that voltage.

### 13.1.2 The Sampling Process

The analog voltage is converted to a digital representation using an analog-to-digital converter such as the flash converters described in Chapter 2. This device performs two functions simultaneously: sampling and quantization. Although these functions occur together, we will discuss them separately since they have different effects.

Figure 13.4(a) shows an analog voltage represented as a function of time, and Figure 13.4(b) shows the same voltage after sampling. The sampling process can be considered a mechanism for approximating the waveform. At discrete times, the waveform is interrogated and that value remembered until the next sampling time. The sampled analog waveform thus consists of a series of *steps*, with constant values between the steps. The conditions under which this process results in an accurate approximation of the waveform will be discussed later.

### Resolution

The *resolution* of a system is determined in large part by the sampling process. The number of samples on a single line defines the

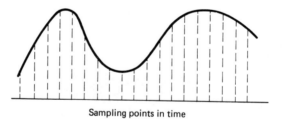

Sampling points in time

**Figure 13.4(a)**    Analog signal.

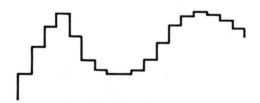

**Figure 13.4(b)**    Sampled analog signal.

*horizontal resolution* of the system. Similarly, the number of lines defines the *vertical resolution*. This is demonstrated by a comparison of American and European television in which the greater vertical resolution of the European picture, with the standard of 625 lines, is obvious to the viewer.

One common sampling rate is 100 ns (nano seconds) per pixel. Such a sampling rate is easily derived from a 10-MHz clock and results in just over 512 samples on each line. The number 512 is very convenient from a hardware point of view since it is a power of 2. Using a sampling rate of 103 ns/pixel gives 512 samples on exactly one full line of video.

### Dynamic Range

Once an analog signal has been sampled, it is converted to digital form by a process known as *quantization* as shown in Figure 13.5. The digital representation of a signal can have only a finite number of possible values, defined by the number of bits in the output word. For example, video signals are often encoded to 8 bits of accuracy, thus allowing a signal to be represented as one of a possible 256 values. A larger number of bits allows a signal to be represented to a greater degree of accuracy.

The accuracy (number of bits) of the digital representation is often referred to as the *dynamic range* of the imaging system. We must caution the reader that the meaning of the term dynamic range sometimes varies. An alternative definition specifies the dynamic range as the range of input signals over which a camera successfully operates. For example, if we open the iris on the camera, we may be said to alter the dynamic range of the camera. Both meanings are accepted and are in common use, but they differ according to their contexts.

Thus, we conclude that the digital image is "discrete in space and discrete in value." We also observe that there is a one-to-one relationship between time and space. That is, if we refer to the *sampling time*,

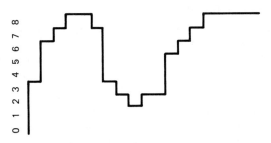

**Figure 13.5**    Quantized sampled signal.

we must speak of it relative to the top-of-picture signal (vertical blanking). That timing relationship identifies a unique position on the screen.

### The Sampling Theorem

In 1948, Claude Shannon derived the *sampling theorem*. This theorem states, simply, that if an analog signal is filtered by an ideal low-pass filter with cutoff at a frequency $f$, then that (filtered) signal can be exactly reconstructed if it is sampled using a sampling rate greater than or equal to $2f$. Said another way, if we wish to sample and store an analog signal and to be able to reconstruct that signal exactly from the sampled version, our sampling rate must be at least equal to twice the highest frequency in the signal.

The sampling theorem addresses only the effects of errors due to the sampling process; it says nothing about the effects of quantization.

Black-and-white television systems are generally designed to have a 5-MHz bandwidth. Thus, a sampling rate of 10-MHz is required to completely recover the original signal. In fact, the sampling rate of 103 ns which we proposed earlier is very close to this rate. We chose 512 samples per line rather arbitrarily in the earlier discussion, presumably because it was a power of 2. Now, we can see, however, that 512 pixels is not only a power of 2, but it also results naturally from the application of the sampling theorem to the video signal.

The sampling theorem appears to lead to a fascinating contradiction. If we undersample the video signal, say, with 100 points per line, or even lower, and then reconstruct the image, when you and I look at it, we can still recognize the contents of the scene. How can this be? The sampling theorem seems to say that we must sample at 512 (or more) points per line.

In fact, there is no contradiction at all. The sampling theorem says that we must sample at twice the bandwidth to be able to reconstruct the image *without distortion*. Images are highly redundant in their information content, and the human brain is an excellent image-interpolating machine.

### Example 13.1  Effects of Sampling

Show the effects of undersampling a television image in both the horizontal and vertical directions. Assume an initial image of 512 lines of 512 pixels each. Show the results of 2 : 1, 4 : 1, and 8 : 1 undersampling. Determine the clock rate required to produce such images, and the memory required to store them.

**Solution:**   Figure 13.6 shows the results of the three cases. To achieve 2 : 1 undersampling horizontally, we use a clock period of $103 \times 2 = 206$ ns per clock. To achieve such undersampling vertically, we store only every other line. 64K bytes of memory are required (assuming that 1 byte is used to store 1 pixel).

**Figure 13.6(a)**  Image  represented  by
using 256 × 256 pixels.

**Figure 13.6(b)**  Image  represented  by
using 128 × 128 pixels.

**Figure 13.6(c)**  Image  represented  by
using 64 × 64 pixels.

For 4:1 undersampling, the clock period = 412 ns, and we store every fourth line. 16K bytes of memory are required.

For 8:1 undersampling, the clock period = 1.8 μs, every eighth line is stored, and 4K are required.

*Quantization error* is the term used to refer to the fact that information is lost whenever the continuously valued analog signal is partitioned into discrete ranges by the limited dynamic range of the A/D converter. Quantization error is observed in reconstructed images either as random noise effects or as *contouring*. Figure 13.7(a) shows an image that has been quantized to 16 grey levels; Figure 13.7(b) shows the same image quantized to 8 levels.

Experiments have shown that at a given light level, the human eye can discern only about 30 grey levels. Of course, with an overall change in brightness, the eye undergoes *dark adaptation*, a chemical process, and the iris opens; therefore, *which* 30 shades of grey are distinguishable may vary from one environment to another.

**Figure 13.7(a)** Sixteen grey levels.

**Figure 13.7(b)** Eight grey levels.

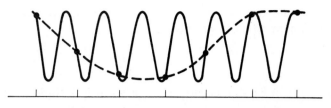

**Figure 13.8**    An analog sine wave sampled at too low a rate.

Thirty shades of grey would seem to indicate that 5 bits is all the dynamic range needed. Most systems, however, use 8 bits, which allows a limited emulation of the effects of the iris. Furthermore, 8-bit memories are conveniently available.

### Aliasing

*Aliasing* is the phenomenon that occurs when the requirements of the sampling theorem are not met. This may be best explained by considering Figure 13.8, which shows a sine wave being sampled at a sampling rate only slightly less than its own frequency. The data acquisition system is unable to correctly reconstruct the original signal because it "thinks" that it is sampling a signal much lower in frequency, as shown by the dotted line.

**Example 13.2  Effects of Quantization**

Show the effects of quantizing a $512 \times 512$ image using 16 shades of grey and 8 shades of grey. Determine the memory required for each.

**Solution:**    See Figure 13.7.  Sixteen shades of grey can be encoded into 4 bits. Therefore $512 \times 512 \times 4 = 1$ megabits of memory is required.  This would probably be stored as 2 pixels per byte.

Eight shades of grey can be encoded into 3 bits.  Therefore, $512 \times 512 \times 3 = 786,432$ bits.  However, it would be difficult to store 3-bit/pixels in a byte-addressed memory.  The most convenient means would probably be a 4-bit/pixel, resulting in the same memory requirements as the 16-shade case.

## 13.2 IMAGE PROCESSING FUNCTIONS

In Section 13.1, we discussed how digital images could be acquired, stored, and represented.  Once this proces is completed, and we have a digital representation of our image stored in the memory of our computer, we can begin to operate on that image and use it in industrial applications.  There are three generic classes of operations that we could ask the computer to perform on images: enhancement, restoration, and analysis.  Analysis is the emphasis of the remainder of this chapter, and we will, therefore, touch only on enhancement and restoration here to a degree sufficient to define the terms.

### Enhancement

Image enhancement could be most simply defined as the science of making images "look better." Generally, this means looking better from the point of view of a human observer, although enhancement techniques may also be used to preprocess an image prior to analyzing it.

In general, the computer does not need to have any knowledge of the contents of the scene to enhance it. As an example, we will describe what is probably the simplest form of enhancement, grey scale stretching, or contrast stretching.

First, we (that is, the computer) make a pass over the image, memorizing the largest intensity value in the image, and the smallest. Let us refer to them as $I_{max}$ and $I_{min}$. If the maximum possible value is 255 (for 8-bit images) and the smallest possible value is 0, we can stretch the contrast by updating each pixel value in the following way:

$$I_{new} = \frac{(I_{old} - I_{min}) \cdot 255}{(I_{max} - I_{min})} \tag{13.3}$$

Thus, the new image will have a maximum value of 255 and a minimum value of 0, effectively utilizing the total dynamic range of the output (viewing) system. Similar techniques are also useful on input to normalize images to compensate for variations in lighting.

Many more sophisticated techniques exist for enhancing images, including histogram modification and edge enhancement (see Pratt, 1978; Gonzalez and Wintz, 1977).

### Restoration

In the process of acquiring images, distortion always occurs (Chien and Snyder, 1975). There are many sources of distortion, including

*Vignetting:* Lenses transmit more efficiently near the center than near the outside, resulting in images that are darker near the outside.

*Parabolic distortion:* Electron beam efficiency is greater if the beam is normal to the faceplate, resulting in exactly the same effects as vignetting.

*Blooming:* Too much light in a single spot on the faceplate allows the accumulated charges to diffuse outward during a single scan time, resulting in bright spots that seem too large.

*Lag:* The electron beam is not 100 percent efficient, leaving some accumulated charge not neutralized after each scan, resulting in a "ghost" image when the scene changes.

*Motion blur:* If image motion is more rapid than the scan time, the image of the same point may be smeared over several pixels.

*Geometric distortion:* Lenses are not perfect, nor are faceplates, and the resulting image may be distorted geometrically. The image of a perfect circle, for example, may not be a perfect circle.

*Blur:* Improperly focused optics or an improperly focused electron beam can result in an image in which edges are not as sharp as they could be. In fact, in almost all systems of high resolution (e.g., 512 X 512), edges are never perfect step functions.

If we known the exact mathematical form of these distortions, we could write

$$g(x, y) = D[f(x, y)] \qquad (13.4a)$$

where $f(x, y)$ is the "true" undistorted image, $D$ is a distortion operator, and $g(x, y)$ is the measured image. If the $D$ operator has an inverse, then we could recover the original image from the distorted one by simply computing

$$f(x, y) = D^{-1}[g(x, y)]. \qquad (13.4b)$$

Of course, in general, $D$ is extremely complex and may not have an inverse. The field of research that studies such distortion functions and attempts to find inverse operators is called *image restoration*. In most industrial applications, restoration techniques are not used explicitly or rigorously. However, corrections for geometric distortion (warping) are routinely done (see Pratt, 1978; Gonzalez and Wintz, 1977; Andrews and Hart, 1977).

## 13.3 IMAGE ACQUISITION HARDWARE

There are many ways in which to acquire and access a digital image. Here, we will describe only one technique, which allows the acquisition of a single frame of video data in real frame.

### 13.3.1 The Frame Buffer

Figure 13.9 shows a block diagram of a digital data acquisition system designed to acquire and store one frame of video data, quantized

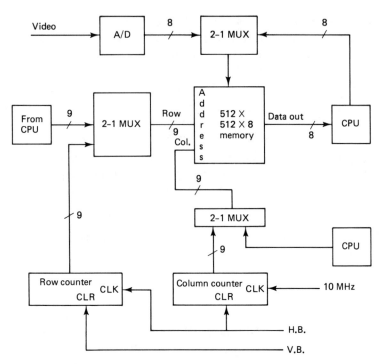

**Figure 13.9**   A video data acquisition system.

to 8 bits per pixel.  The memory in this system, referred to as a *frame buffer,* will store a 512 X 512 picture and will acquire it in 33 ms.*

The fundamental concept behind the design is the use of counters which count in synchronism with the motion of the electron beam.  The 9-bit column counter counts from 0 to 511 at a 100-ns/count rate. The column counter is cleared by the horizontal blanking pulse (HB). Hence, a zero in this counter indicates that the beam is at the left-hand end of a scan line.

Although the horizontal blanking pulse clears the column counter, it is used to increment the row counter.  Thus, the row counter keeps track of which row the electron beam is currently on.  The vertical blanking pulse (VB), which signifies that the beam is at the top of the picture, is used to clear the row counter.

The combination of the row and column counters composes the 18-bit address to the memory.  Thus, for each position on the screen, there exists a unique address.

---

*In this simple design, we are assuming interlaced scanning.

To make use of the information stored in the frame buffer, the computer must be able to read it. This is accomplished via 2-1 multiplexers, one on the row address bus and one on the column address bus. During a computer access, the 18-bit address is provided by the computer rather than coming from the counters. To prevent erroneous addressing, the memory WRITE operation of data from the A/D converter must be disabled while the computer is reading.

This rather straightforward design provides data at the video rate. In the past, such high-speed configurations have not been popular due to the cost of high-speed A/D converters and memory. With the rapid decline in costs of both items, this *frame grabber* technique has now replaced virtually all the other, lower-speed techniques.

Another factor influencing the popularity of frame grabbers is the fact that the same hardware can provide raster scan graphics capabilities. In a graphics application, the memory is read in synchronism with the scan. The data are converted to analog video by a D/A converter and are transferred from there to a TV monitor. Consequently, to convert a raster scan graphics system to a frame grabber requires little more than addition of an A/D converter.

**The Pixel Array**

In this section, we have discussed a video acquisition system capable of sampling the video signal at a rate of one sample every 100 ns, resulting in an array of 512 samples on each line and 512 lines. Once stored in a digital memory, this array may be accessed as a conventional two-dimensional array of numbers. Each number is referred to as a *pixel*, short for "picture cell." (The term *pel*, representing "picture element," is sometimes used in the same way, especially in Europe.)

Typically a single pixel is represented by one 8-bit byte. The most important observation to be made at this point with respect to the pixel array is its size. A typical 512 X 512 array requires a quarter of a million bytes of memory. At one time, this would have been a prohibitive amount of memory. Memory cost is no longer a significant restriction, but using a serial computer on such a massive amount of data is still extremely time consuming. For this reason, much of the ongoing research in vision today is directed toward the development of algorithms that can make use of parallelism (Fu and Ichikawa, 1982).

In addition to frame grabbers, several manufacturers are now marketing frame buffer systems augmented with other special high-speed hardware for performing operations on the digital image. In the next section, we will examine a few of these operations.

## 13.4 SEGMENTATION

In many robot vision applications, the set of possible objects in the scene is quite limited. For example, if the camera is viewing a conveyer, there may be only one type of part which appears, and the vision task could be to determine the position and orientation of the part. In other applications, the part being viewed may be one of a small set of possible parts, and the objective is to both locate and identify each part.* Finally, the camera may be used to inspect parts for quality control.

In this section, we will assume that the parts are fairly simple and can be characterized by their two-dimensional projections, as provided by a single camera view. Furthermore, we will assume that the shape is adequate to characterize the objects. That is, color or variation in brightness is not required. We will first consider dividing the picture into connected regions.

A *segmentation* of a picture is a partitioning into connected regions, where each region is homogeneous in some sense and is identified by a unique label. For example, in Figure 13.10, region 1 is identified as the background. Although region 4 is really background also, it is labeled as a separate region since it is not connected to region 1. While there are several ways to perform segmentation, we will discuss only one here.

### 13.4.1 Segmentation by Thresholding

In applications where grey scale is not important, we can segment a picture into "objects" and "background" by simply choosing a threshold in brightness. We define any regions whose brightness is above the threshold as *object* and all below the threshold as *background*.

There are several different ways to choose thresholds, ranging from trivially simple techniques to quite sophisticated methods. As the sophistication of the technique increases, performance improves at the cost of increased computational complexity.

Probably the most important factor to note is the *local nature of thresholding*. That is, a single threshold is almost never appropriate for an entire scene. It is nearly always the local contrast between object and background that contains the relevant information. Since camera sensitivity drops off from the center of the picture to the edges due to parabolic distortion and/or vignetting, it is useless to attempt to establish a global threshold. A dramatic example of this effect can be seen in an

---

*A special case of this application occurs when only one type of part comes down the line, but that part has substantial three-dimensional structure and may be resting in one of several possible *stable states*. In this case, the view corresponding to each stable state may be treated as the view of a different object.

**Figure 13.10(a)**  A  picture  with  two foreground regions.

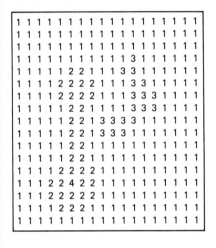

**Figure 13.10(b)**  A  segmentation  and labeling of the picture in Figure 13.10(a). Region 4 is a hole in region 2.

image of a chess board, in which the white squares at the corners are actually darker than the black squares in the center.

Effects such as parabolic distortion and vignetting are quite predictable and easy to correct. In fact, off-the-shelf hardware is available for just such applications. It is more difficult, however, to predict and correct effects of nonuniform ambient illumination, such as sunlight through a window, which changes radically over the day.

Since a single threshold cannot provide sufficient performance, we must choose local thresholds. The most common approach is called *block thresholding*, which the picture is partitioned into rectangular blocks and different thresholds are used on each block. Typical block sizes are 32 × 32 or 64 × 64 for 512 × 512 images. The block is first analyzed and a threshold is chosen; then that block of the image is thresholded using the results of the analysis.

### Choosing a Threshold

The simplest strategy for choosing a threshold is to average the intensity over the block and choose $i_{avg} + \Delta i$ as the threshold, where $\Delta i$ is some small increment, such as 5 out of 256 grey levels. Such a simple thresholding scheme can have surprisingly good results (Page, Snyder, and Rajala, 1983).

However, when the simpler schemes fail, one is forced to move to more sophisticated techniques, such as thresholding based on histogram analysis. Before we describe this technique, we will first define a histogram.

The *histogram* $h(i)$ of an image $i(x, y)$ is a function of the permissible intensity values. In a typical imaging system, intensity takes on values between 00 (black) and $FF_{16}$ (white). A graph that shows, for each grey level, the number of times that level occurs in the image is called the histogram of the image. Figure 13.11 shows a typical histogram for an image of black parts on a white conveyor.

In Figure 13.11 we note two distinct peaks, one at grey level 3, almost pure black, and one at grey level 193, bright white. With the exception of noise pixels, every point in the image belongs to one of these regions. A good threshold, then, is anywhere between the two peaks.

Histograms are seldom as "nice" as the one in Figure 13.11 and some additional processing is generally needed (Chew and Kaneko, 1972; Rosenfeld and Kak, 1976). However, the philosophy of his-

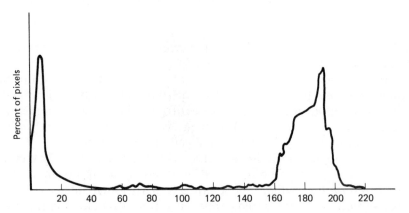

**Figure 13.11**   Histogram of a picture with a mixture of pure black and pure white.

togram-based thresholding is the same: find two peaks in the histogram and choose the threshold to be between them.

In general, different thresholds are used in different areas of the picture. In many industrial environments, the lighting may be extremely well controlled. With such control, the best thresholds will be constant over time and may be chosen interactively during system set up. In other, more variable, situations, the computer may be required to analyze the distribution of grey levels over the area of interest to choose an appropriate threshold.

### Region Labeling

Let us assume, for now, that a good threshold has been chosen and that our picture has been partitioned into regions of pure black and pure white, as shown in Figure 13.10(a). The production of a segmented picture such as Figure 13.10(b) requires an analysis of connectedness. That is, a pixel is in region $i$ if it is above threshold and is adjacent to a pixel in region $i$. Since regions may curve and fork, the analysis cannot be as simple as starting at the top and marking connected pixels going down. Instead a more sophisticated technique is needed.

One such algorithm is known as *region growing*. It utilizes a label memory corresponding to the frame buffer just as Figure 13.10(b) corresponds to 13.10(a). In this description, we will refer to "black" pixels as object and "white" as background.

Initially, each cell in the label memory $M$ is set to zero. We will refer to the picture memory as $P$. Thus, the grey scale of a point with coordinates $\langle x, y \rangle$ is $P(x, y)$, and the labeling operation can be written $M(x, y) \leftarrow N$ for some label number $N$.

### Algorithm Grow

This algorithm implements region growing by using a push down stack on which to temporarily keep the coordinates of pixels in the region.

1. Find an unlabeled black pixel; that is, $M(x, y) = 0$. Choose a new label number for this region, call it $N$. If all pixels have been labeled, stop.

2. If $P(x - 1, y)$ is black and $M(x - 1, y) = 0$, push $\langle x - 1, y \rangle$ onto the stack.

    If $P(x + 1, y)$ is black and $M(x + 1, y) = 0$, push $\langle x + 1, y \rangle$ onto the stack.

If $P(x, y - 1)$ is black and $M(x, y - 1) = 0$, push $\langle x, y - 1 \rangle$ onto the stack.

If $P(x, y + 1)$ is black and $M(x, y + 1) = 0$, push $\langle x, y + 1 \rangle$ onto the stack.

3. $M(x, y) \leftarrow N$.
4. Choose a new $\langle x, y \rangle$ by popping the stack.
5. If the stack is empty, go to 1, else go to 2.

This labeling operation results in a set of connected regions, each assigned a unique label number. To find the region to which any given pixel belongs, the computer has only to interrogate the corresponding location in the $M$ memory and read the region number.

### Example 13.3  Applying Region Growing

The figure below shows a 4 × 7 array of pixels. Assume the initial value of $\langle x, y \rangle$ is $\langle 2, 4 \rangle$. Apply algorithm "grow" and show the contents of the stack and $M$ each time step 3 is executed. Let the initial value of $N$ be 1.

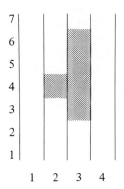

**Solution:**
Pass 1: Immediately after execution of step 3.

Stack: $\langle 3, 4 \rangle \leftarrow$ top

$$M = \begin{array}{c|c|c|c|c|} 7 & 0 & 0 & 0 & 0 \\ \hline 6 & 0 & 0 & 0 & 0 \\ \hline 5 & 0 & 0 & 0 & 0 \\ \hline 4 & 0 & 1 & 0 & 0 \\ \hline 3 & 0 & 0 & 0 & 0 \\ \hline 2 & 0 & 0 & 0 & 0 \\ \hline 1 & 0 & 0 & 0 & 0 \\ \end{array}$$
$$\quad\quad 1\ \ 2\ \ 3\ \ 4$$

Pass 2:

Stack: ⟨3, 5⟩ ← top
⟨3, 3⟩

$$M = \begin{array}{c|c|c|c|c|} 7 & 0 & 0 & 0 & 0 \\ 6 & 0 & 0 & 0 & 0 \\ 5 & 0 & 0 & 0 & 0 \\ 4 & 0 & 1 & 1 & 0 \\ 3 & 0 & 0 & 0 & 0 \\ 2 & 0 & 0 & 0 & 0 \\ 1 & 0 & 0 & 0 & 0 \\ \hline & 1 & 2 & 3 & 4 \end{array}$$

Pass 3:

Stack: ⟨3, 6⟩ ← top
⟨3, 3⟩

$$M = \begin{array}{c|c|c|c|c|} 7 & 0 & 0 & 0 & 0 \\ 6 & 0 & 0 & 0 & 0 \\ 5 & 0 & 0 & 1 & 0 \\ 4 & 0 & 1 & 1 & 0 \\ 3 & 0 & 0 & 0 & 0 \\ 2 & 0 & 0 & 0 & 0 \\ 1 & 0 & 0 & 0 & 0 \\ \hline & 1 & 2 & 3 & 4 \end{array}$$

Pass 4:

Stack: ⟨3, 3⟩ ← top

$$M = \begin{array}{c|c|c|c|c|} 7 & 0 & 0 & 0 & 0 \\ 6 & 0 & 0 & 1 & 0 \\ 5 & 0 & 0 & 1 & 0 \\ 4 & 0 & 1 & 1 & 0 \\ 3 & 0 & 0 & 0 & 0 \\ 2 & 0 & 0 & 0 & 0 \\ 1 & 0 & 0 & 0 & 0 \\ \hline & 1 & 2 & 3 & 4 \end{array}$$

Pass 5:

Stack: Empty

$$M = \begin{array}{c|cccc|} 7 & 0 & 0 & 0 & 0 \\ 6 & 0 & 0 & 1 & 0 \\ 5 & 0 & 0 & 1 & 0 \\ 4 & 0 & 1 & 1 & 0 \\ 3 & 0 & 0 & 1 & 0 \\ 2 & 0 & 0 & 0 & 0 \\ 1 & 0 & 0 & 0 & 0 \\ \hline & 1 & 2 & 3 & 4 \end{array}$$

This region-growing algorithm is just one of several strategies for performing *connected component analysis*. Other strategies exist which are faster than the one described, including some that run at raster scan rates (Snyder and Savage, 1982). We will now consider techniques for making use of this information.

## 13.5 SHAPE DESCRIPTORS

In the process of generating the segmented version of a picture, the computer performs a region-growing operation that acts on each pixel in the region. In so doing, the computer can easily keep track of the area. Area is one of many features that can help us to distinguish one type of object from another. For example, the image of a connecting rod typically occupies more area (more black pixels) than does the image of a valve. Thus, by measuring the area of a region, we may discern the type of object.

In this section we will present a few of the many other features which may be used to characterize regions (Ballard and Brown, 1983).

### 13.5.1 Features

*Average grey value:* In the case of black and white "silhouette" pictures, this is simple to compute.

*Maximum grey value:* Is straightforward to compute.

*Minimum grey value:* Is straightforward to compute.

*Area:* Comes directly from the region-growing algorithm.

*Perimeter:* Several different definitions exist. Probably the simplest is a count of all pixels in the region that are adjacent to a pixel not in the region.

*Diameter:* The diameter describes the maximum chord—the distance between those two points on the boundary of the region whose mutual distance is maximum (Snyder and Tang, 1980; Shamos, 1975).

*Thinness:* Two definitions for thinness exist: $T_A = P^2/A$ measures the ratio of the squared perimeter to the area; $T_B = D/A$ measures the ratio of the diameter to the area.  Figure 13.12 compares these two measurements on example regions.

*Center of gravity:* The $x$ and $y$ coordinates of the center of gravity may be written

$$m_x = \frac{1}{N}\sum x$$

$$m_y = \frac{1}{N}\sum y$$

for all points in a region with $N$ points.

*X-Y aspect ratio:* See Figure 13.13(a).  The aspect ratio is the length/ width ratio of the bounding rectangle.  This is simple to compute.

*Minimum aspect ratio:* See Figure 13.13(b).  Again, a length/width, but much more computation is required to find the minimum such rectangle.

*Moments:* A moment of order $p + q$ may by defined on a region as

$$m_{xy} = \sum x^p y^q$$

This definition assumes that the region is uniform in grey value and that grey value is arbitrarily set to 1.  The area is then $m_{00}$, and we

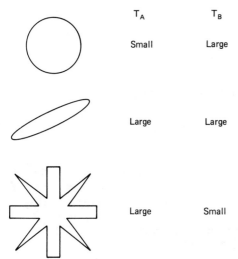

|       | $T_A$ | $T_B$ |
|-------|-------|-------|
|       | Small | Large |
|       | Large | Large |
|       | Large | Small |

**Figure 13.12** Results of applying two different thinness measurements on various regions.

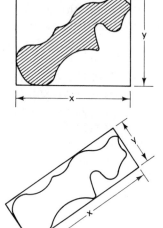

**Figure 13.13(a)**  $y/x$ is the aspect ratio using one definition, with horizontal and vertical sides to the bounding rectangle. This definition is very sensitive to rotations.

**Figure 13.13(b)**  $y/x$ is the minimum aspect ratio. This definition is invariant to rotation.

find that the center of gravity is

$$m_x = \frac{m_{10}}{m_{00}} \qquad m_y = \frac{m_{01}}{m_{00}}$$

We can now define as many features as we wish by choosing higher orders of moments or combinations thereof. Of particular interest are the *invariant moments* (Gonzalez and Wintz, 1977). Those moments have the characteristics that they are invariant to translation, rotation, and scale change, which means that we get the same number, even though the image may be moved, rotated, or zoomed. The first invariant moment is

$$\frac{m_{00}m_{20} - m_{10}^2 + m_{00}m_{02} - m_{01}^2}{m_{00}^3}$$

There are six others (Gonzalez and Wintz, 1977).

*Convex discrepancy:* If one were to stretch a rubber band around a given region, the region that would result would be the *convex hull*

**Figure 13.14**  Convex hull of a region. The shaded area is the convex discrepancy.

(Figure 13.14).   The difference in area between a region and its convex hull is the convex discrepancy.   See Shamos (1975) for fast algorithms for computing the convex hull.

*Number of holes:* One final feature that is very descriptive and reasonably easy to compute is the number of holes in a region.

In this section, several features were defined that could be used to quantify the shape of a region. Some, like the moments, are easy measurements to make.   Others, such as the diameter or the convex discrepancy, require development of fairly sophisticated algorithms to avoid extremely long computation times.   Space does not permit a discussion of those algorithms here, but the reader may find adequate direction in the sources cited in the reference list.

## 13.6 USING SHAPE DESCRIPTORS

The features described in the previous section can be used in many different ways to identify, locate, and orient parts. A thorough discussion of those techniques is the basis of entire books.   In this section we simply introduce the reader to some of the basic concepts of machine vision and to whet his or her intellectual appetite for more knowledge about this potentially very productive field.   For that reason, we will consider only the problem of recognizing parts and leave the issues of location and orientation for another text. As before, we consider only parts which can be recognized by their two-dimensional silhouettes.

We define a *feature vector* $\mathbf{x}$ as an ordered $d$-tuple $\langle x_1 x_2 \cdots x_d \rangle$ in which each element is a scalar feature, as discussed in Section 13.5.

The *event* $w_i$ means that the object being viewed by the camera belongs to class $w_i (i = 1 \cdots c)$. Typical classes of objects would be valves, rods, bolts, covers, and so on.

$P(w_i)$ is the probability that the object being viewed belongs to class $i$, before any measurements have been made. $P(w_i)$ is the a priori probability.

Given a measurement $\mathbf{x}$ made on the object, we can define the conditional probability $P(w_i|\mathbf{x})$ that the object is in class $i$, given that the measurement vector has value $\mathbf{x}$.

Finally, the *conditional probability density* $p(\mathbf{x}|w_i)$ is a function of the feature that represents the probability that a member of class $w_i$ will have feature value $\mathbf{x}$. Note that a different function $p(\mathbf{x}|w)$ exists for each different $w$.   Figure 13.15 shows the probability density functions for one feature and different classes. Since $\mathbf{x}$ is a vector quantity, $p(\mathbf{x}|w)$ is typically a vector function.   For simplicity, Figure 13.15 has

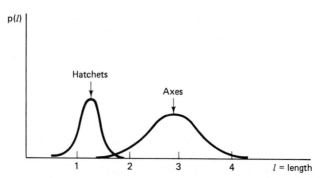

**Figure 13.15**   Probability density function for length of hatchets and axes. From this graph, we see that an object of length 2 is more likely to be an axe than a hatchet.

treated only one of the components of the feature vector, so that the corresponding density functions may be shown as graphs of only two dimensions.

Now, we can state *Bayes' rule*, which relates these probabilities.

$$P(w_i|\mathbf{x}) = \frac{p(\mathbf{x}|w_i)\,P(w_i)}{\sum\limits_{j} p(\mathbf{x}|w_j)\,P(w_j)} \tag{13.5}$$

This equation relates the conditional probability of class $w_i$ (which is what we want) to the conditional density of the measurement (which we know from past experience) and includes the a priori probabilities of each class (also known from prior experience).

We will assume all the classes are equally likely and, therefore, simply ignore $P(w_i)$.  Furthermore, we observe that the denominator is a normalizing factor and does not help us to distinguish one class from another.  Hence we can define a decision rule:

Decide class $w_i$ if, for all $j \neq i$

$$p(\mathbf{x}|w_i) > p(\mathbf{x}|w_j) \tag{13.6}$$

Exactly how to determine and represent $p(\mathbf{x}|w_j)$ can be a subject unto itself.  In the most straightforward approach, it is determined experimentally and is stored in a tabular form.  While convenient for scalars, this approach is awkward for $\mathbf{x}$ vectors of high dimensionality. In these cases, we usually approximate the experimental $p(\mathbf{x}|w)$ with a continuous function.  A Gaussian function is often chosen because of its mathematical simplicity and because the Gaussian approximates naturally occurring densities quite well.

In the Gaussian approximation, we use

$$p(\mathbf{x}|w_i) = \frac{1}{(2\pi)^{d/2} \, |C_i|^{1/2}} \exp\left[ -\frac{1}{2} (\mathbf{x} - \boldsymbol{\mu}_i)^\mathsf{T} \, C_i^{-1} (\mathbf{x} - \boldsymbol{\mu}_i) \right]$$

(13.7)

where $d$ is the dimension of the vector $\mathbf{x}$. $\boldsymbol{\mu}_i$ is the mean vector of a *training set* $\mathbf{X}_i$ of samples, all known to belong to class $i$, and $N_i$ is the number of samples in $\mathbf{X}_i$:

$$\boldsymbol{\mu}_i = \frac{1}{N_i} \sum_{\mathbf{x} \in \mathbf{X}_i} \mathbf{x}$$

(13.8)

The covariance matrix $C_i$ is determined, using that Gaussian assumption, by

$$C_i = \frac{1}{N_i} \sum_{\mathbf{x} \in \mathbf{X}_i} (\mathbf{x} - \boldsymbol{\mu}_i)(\mathbf{x} - \boldsymbol{\mu}_i)^\mathsf{T}$$

(13.9)

With this introduction, we can define a strategy developing a statistical pattern recognition system for recognizing parts.

1. Define features and develop algorithms for measuring features.
2. Choose a large set of objects of type A and measure the features, determine mean and variance for the features for class A. Repeat for classes B, C, and so on.
3. Given an unknown $x$, a part on the conveyor, measure the feature vector $\mathbf{x}$, and apply Eq. 13.7 once for each class. The application returning the highest value is the most likely class.

This discussion has, of course, glossed over most of the field of statistical pattern recognition. Many books are available which go into more detail, including Duda and Hart (1973), Ballard and Brown (1983).

**Example 13.4:  Object Recognition Using Shape Descriptors**

An unknown object is measured by a vision system to have a length of 2.5 units.

A large number of flanges, all of the same type, have been measured and found to have an average (mean) length of 2.1 units and a variance in length of 0.8. Similarly, a large number of gaskets have been measured and found to have an average length of 2.8 units and a variance in length of 1.3.

Is the unknown most likely a flange or a gasket?

**Solution:**    Using the Gaussian assumption described in Eq. 13.7, we find

$$p(2.5|\text{flange}) = \frac{1}{\sqrt{2\pi(0.8)}} \exp \frac{-(2.5 - 2.1)^2}{2 \cdot (0.8)} = 0.404$$

$$p(2.5\,|\,\text{gasket}) = \frac{1}{\sqrt{2\pi(1.3)}}\,\exp\frac{-(2.5-2.8)^2}{2\cdot1.3} = 0.338$$

Using Eq. 13.6, we make the decision that the unknown is more likely to be a flange, since that decision has the higher probability of being correct. However, two observations are in order. First, the two results were very close; therefore, our confidence in the accuracy of our decision is not very high. We need to make another measurement, perhaps the perimeter of the object image, to improve our decision-making capabilities. Second, this analysis has assumed that the Gaussian distribution modeled in Eq. 13.7 accurately reflects the shape of the probability density function. This assumption may not be correct for the length of flanges and gaskets and should be tested before being put into operation.

## 13.7 STRUCTURED ILLUMINATION

For a number of years, popular fantasy painted the picture of the robot with human vision or better as being just around the corner. Today, we know better. Computer vision is a very difficult problem that is not likely to be solved in general in the foreseeable future. However, in the industrial environment, general-purpose vision is not required and probably not even desired. We need only to locate and/or identify a very restricted set of objects. Furthermore, in the industrial environment, we have one other tremendous advantage: we can control the lighting.

The strategy that takes advantage of control of lighting to make the vision problem easier is called *structured illumination*, and it takes many forms. In this section, we will provide three examples, each of which uses controlled lighting to considerable advantage to solve a different problem. One technique eliminates problems with object reflectivity, a second finds objects by triangulation, and a third uses controlled lighting for inspection.

### 13.7.1 Silhouetting

Many industrial parts have the property that they are uniquely identifiable by their silhouette. That is, neither three-dimensional nor grey scale information is required for the part to be uniquely identified and its orientation determined. In these cases, particularly simple techniques exist for extracting and processing the object's silhouette.

For example, in General Motors' CONSIGHT system (Ward et al., 1979) a linear sensor array is used rather than a two-dimensional camera. Such arrays typically have more resolution per line than a camera; 2000 pixels on a line is not uncommon. To form a two-dimensional picture, either the part or the camera may be moved or a mirror may move the image of the part. In CONSIGHT, the part is moved, a trivial task since the sensor is focused on a conveyor.

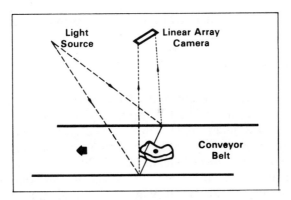

**Figure 13.16**    (Ward) The camera is positioned to image a line across the conveyor belt. (*Robotics Today, used with permission*)

The structured illumination concept applies to CONSIGHT in the way in which the part is illuminated. A narrow stripe of light is focused at an angle on the conveyor at exactly the point of focus of the sensor. With an empty conveyor, the sensor detects a bright bar. When a part moves into the field of view, however, the stripe of light shines on the part (Figure 13.16). Since the part has depth and the light is at an angle, the light stripe is displaced horizontally and the sensor detects darkness. By taking a series of stripe images as the part moves, the system can build up a profile of the part.

The system as implemented (Figure 13.17) actually uses two light projectors. This eliminates the problem of shadows being erroneously identified as parts.

The reflectivity of parts, especially metallic ones, is one of the most difficult problems in image analysis. To see that this is a problem, we should recognize that a shiny metallic part is really a mirror. Consider the intrinsic difficulty in analyzing the image of a mirror. Since the CONSIGHT system makes use of only the displacement of the light

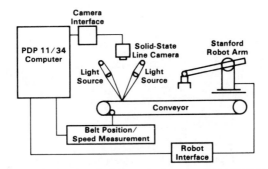

**Figure 13.17**    (Ward) Schematic diagram of CONSIGHT.

stripe, it has the additional advantage that it is independent of the surface reflectivity of the part.

### 13.7.2 Range Triangles

In many systems, the concept of the range triangle has been used to determine three-dimensional position and orientation of parts. In Page, Snyder, and Rajala (1983), a system is described which determines the position and orientation (often these two terms are lumped together and are referred to as *pose*) of steam turbine blades.

Such blades have shiny metallic surfaces and, therefore, reflect most of the light in undesirable ways. However, the surface scatters enough light back to the camera to detect a horizontal bar, if such a bar is projected on the blades.

Figure 13.18 shows a light source projecting a narrow (0.2 centimeters) stripe of light across a scene containing a blade. The scene detected by the camera is shown in Figure 13.19. The horizontal light stripe appears curved to the camera when it is reflected off the blade. The curved stripe is easy to detect in the image by using some of the thresholding strategies described earlier in this chapter.

The three-dimensional position of the two corners of the blade are determined by making use of a *range triangle*, as shown in Figure 13.20. The camera position is known, as is the position of the light projectors. The angle of the projector with respect to the scene is also known by a calibration procedure. From the camera image, the computer finds the angle $\alpha$, which is all that is needed to determine the distance $d$ to the part.

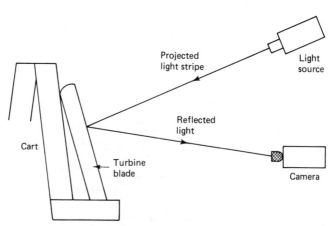

**Figure 13.18**   System layout (side view).

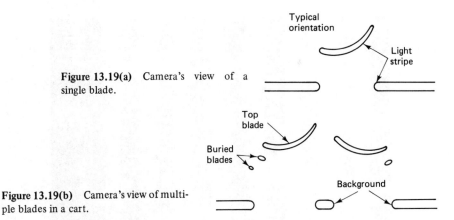

**Figure 13.19(a)**   Camera's   view   of   a
single blade.

**Figure 13.19(b)**   Camera's view of multiple blades in a cart.

Again, the combination of controlled lighting and simple image
processing has provided a reasonable operational industrial vision system.

### 13.7.3 Range Sensors

One of the most exciting prospects of current research in image
analysis is the development of special sensors which explicitly determine a *range image*. By range image we mean a two-dimensional array
of pixels, exactly like a conventional "luminance" image. However, the
number stored in each pixel location represents the Euclidean distance
from the sensor to the surface being viewed, rather than the relative
brightness of the surface at that point. Such sensors may employ various structured lighting strategies, such as those described earlier, or

**Figure 13.20**   Range triangle.   The depth of the point was derived from the following trigonometric relationships:

$$\gamma = 90° - \alpha \qquad \frac{\sin(\beta)}{r} = \frac{\sin(180° - \beta - \gamma)}{h}$$

Recall that $h$ and $\beta$ were known and that $\alpha$ was measured by the camera. Then,

$$r = \frac{[h \times \sin(\beta)]}{\sin(180° - \beta - \gamma)}$$

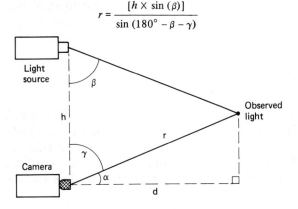

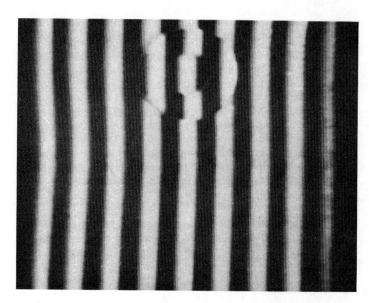

Figure 13.21    (Porter & Mundy) Sinusoidally modulated light. (*Courtesy 1982 IEEE*)

even more sophisticated techniques, such as pulsed lasers. Independent of the internal strategy of the sensor, the result is a range picture. Such a picture gives surface pose information immediately and, thus, can readily provide the three-dimensional information needed by robots for grasping.

Such sensors are still under development, as are techniques for making use of such information. Porter and Mundy (1982) describe one such sensor that uses sinusoidally modulated light patterns (Figure 13.21). The sensor makes use of the difference of two patterns projected at different wavelengths to reduce the undesirable effects of surface reflectivity. The system reported is used in an inspection application to identify small defects on parts.


## 13.8 CONCLUSION

Vision provides the robot with its most flexible and powerful capability. In this chapter we have presented a few of the currently relevant research and development topics in the area of industrial machine vision. With the introduction and terminology provided here, the reader can more easily follow the developments in this exciting and growing field.

## 13.9 SYNOPSIS

### Vocabulary

You should know the definition and application of the following terms:

aliasing
aspect ratio
blanking
block thresholding
blur
composite video
contouring
convex discrepancy
convex hull
decision rule
dynamic range
enhancement
feature
feature vector
frame buffer
frame grabber
frame time
Gaussian density
geometric distortion
histogram
industrial machine vision
invariant moment
lag
moment
motion blur
parabolic distortion
pel
pixel
quantization
range sensor
raster scan
region growing
region labeling
resolution
restoration
sampling
sampling theorem

segmentation
shape descriptor
silhouetting
structured illumination
thinness
thresholding
undersampled
videcon
vignetting
warping

**Notation**

| Symbols | Meanings |
|---|---|
| $i$ | Incident light intensity |
| $t_f$ | Frame time |
| $q$ | Charge |
| $f(x, y)$ | The true image |
| $g(x, y)$ | The image perceived by the sensor system |
| $D$ | Distortion operator |
| $m_{xy}$ | A moment |
| $T_A$ | One definition of thinness |
| $T_B$ | Another definition of thinness |
| $\mathbf{x}$ | A feature vector |
| $P(w_i)$ | Probability of event $w_i$ |
| $p(\mathbf{x}\|w_i)$ | Probability density of making measurement $\mathbf{x}$, given that event $w_i$ has occurred |
| $\boldsymbol{\mu}_i$ | Mean of a Gaussian (normal) distribution |
| $C_i$ | Covariance of a Gaussian distribution |
| $X_i$ | Training set |

## 13.10 REFERENCES

ANDREWS, H. C., and HUNT, B. R. *Digital image restoration.* Englewood Cliffs, NJ.: Prentice-Hall, 1977.

BALLARD, D. H., and BROWN, C. M. *Computer vision.* Englewood Cliffs, N.J.: Prentice-Hall, 1983.

CASTLEMAN, R. R. *Digital image processing.* Englewood Cliffs, N.J.: Prentice-Hall, 1977.

CHIEN, R. T., and SNYDER, W. E. "Hardware for Visual Image Processing." *IEEE Transactions on Circuits and Systems, 22*, June 1975.

CHOW, C. K., and KANEKO, T. Automatic boundary detection of the left ventricle from cineangiograms. *Computers and Biomedical Research*, August 1972.

COOLEY, J. W., and TUKEY, J. W. An algorithm for the machine calculation of complex Fourier series. *Math. of Computation*, Vol. 19, April 1965, pp. 299–301.

DUDA, R. O., and HART, P. E. *Pattern classification and scene analysis*. New York: John Wiley, 1973.

FU, K. S., and ICHIKAWA, T. *Special computer architectures for pattern processing*. Boca Raton, Fl.: CRC Press, 1982.

FU, K. S., ed. *Applications of Pattern Recognition*. Boca Raton, Fl.: CRC Press, 1980.

GONZALEZ, R. C., and WINTZ, P. *Digital image processing*. Reading, Mass.: Addison-Wesley, 1977.

PAGE, N., SNYDER, W. E., and RAJALA, S. A. Turbine blade image processing system. First International Conference on Advanced Software in Robotics, Liège, Belgium, May, 1983.

PORTER, G. B., and MUNDY, J. L. A noncontact profile sensing system for visual inspection. 6th International Conference on Pattern Recognition, Munich, Oct. 1982.

PRATT, W. K. *Digital image processing*. New York: John Wiley, 1978.

ROSENFELD, A., and KAK, A. C. *Digital picture processing*. New York: Academic Press, 1976.

SHAMOS, M. I. Geometric complexity. *Proceedings of the 7th ACM Symposium on Theory of Computing*, May 1975.

SNYDER, W. E., and SAVAGE, C. "Content-Addressable Read/Write Memories for Image Analysis," pp. 963–968. *IEEE Transactions on Computers*, October 1982.

——, and TANG, D. A. "Finding the extrema of a region. *IEEE Transactions on Pattern Analysis and Machine Intelligence*, Vol. 3, pp 266–269, May 1980.

WARD, M. R., ROSSOL, L., HOLLAND, S. W., and DEWAR, R. Consight: An adaptive robot with vision. pp. 26–32. *Robotics Today*, Summer 1979.

## 13.11 PROBLEMS

1. One type of slow scan television produces one frame of 525 lines every second. Determine the sampling rate necessary to produce an image with 525 lines and 256 pixels of video on each line. Assume that blanking occupies 18 percent of of the line time.

2. Assume that you are given a memory which can acquire data at the rate of 1 pixel every 200 ns. Discuss the options you have in trading off dynamic range and resolution, given that you must acquire data at video rates.

3. For the system described in problem 2, design a circuit that will acquire 2-bit pixels and pack them into 8-bit memory words to be read into the memory. Assume that the memory will accept data when it is given a MSTROBE signal, which you must provide.

4. A pattern consisting of 500 vertical black bars on a white background is scanned by a conventional, black and white, NTSC TV camera. The width of the white and black areas is equal, and the array of bars encompasses the entire frame. The TV signal is sampled by a clock scanning at 103 ns per sample. This results in an undersampled signal, resulting in an aliased signal at a lower frequency. What will be this frequency (or the dominant frequency component of the aliased signal). Note: The solution of this problem may require research into, or knowledge of, signal processing literature not included in this text.

5. The equation for contrast stretching (Section 13.2) could be written $I_{new} = (I_{old} - I_{min})\,\alpha$. A picture is determined to have a maximum brightness of 200 and a minimum brightness of 60. Determine $\alpha$. Show that this equation will stretch the contrast to between 0 and 255.

6. Determine the memory capacity required to store a $512 \times 512$ array of 6-bit pixels. Discuss alternative memory organizations.

7. The figure below shows a $7 \times 4$ array of pixels. Determine a "good" threshold to separate black from white pixels. Justify your decision based on a histogram.

| 61 | 55 | 60 | 61 |
|----|----|----|----|
| 62 | 63 | 57 | 58 |
| 56 | 58 | 38 | 32 |
| 55 | 35 | 36 | 37 |
| 37 | 34 | 39 | 35 |
| 40 | 39 | 38 | 39 |
| 35 | 34 | 39 | 40 |

8. A bimodal histogram (that is, a histogram with two peaks) can be modeled as the sum of two Gaussian density functions.
   (a) Simplify Eq. 13.7 for the case where $d = 1$. In this case, the vectors become scalars and multiplication by the inverse of a matrix becomes division by a constant. (You may wish to look up the form of the normal density function.)
   (b) Represent a bimodal histogram as the sum of two such functions.
   (c) Suppose that we have an experimental histogram such as that illustrated in Figure 13.11, which we wish to model analytically by the equation derived in part 2. What are the unknowns? Discuss how one might go about solving for these unknowns.

9. Algorithm *grow* (Section 13.4.1) gives one approach to labeling of connected components. This approach requires two frame buffer memories: one to hold the image and one to hold the labels. In addition, use of the stack is rather

slow.   An alternative algorithm would be to label pixels as they come from the camera, in raster scan order.   That is, a black pixel at $\langle x, y \rangle$ would be assigned the same label as the pixel at $\langle x - 1, y \rangle$ if that pixel were black, or assigned the same label as the pixel at $\langle x, y - 1 \rangle$ if that pixel were black.

(a)  Develop a flow chart for assigning labels in this way

(b)  Discuss any problems which arise.  For example, consider U-shaped regions.

10.  The following figure shows a 4 $\times$ 7 region of an image, containing a black figure on a white background.

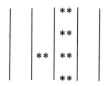

(a)  What is the area of this region?

(b)  What is the diameter of the region. Discuss any ambiguities in the definition.

(c)  What is the perimeter? Propose at least two definitions.

(d)  Determine and compare the two definitions of thinness when applied to this region.

(e)  Determine the center of gravity.

(f)  Find the smallest rectangle that encloses the region.

(g)  Determine the convex hull and convex discrepancy.

11.  Example 13.4 presents a technique for distinguishing between flanges and gaskets using only length.  In this problem, we again distinguish flanges from gaskets, but using both length and weight.  We define

$$\mathbf{x} = \begin{bmatrix} \text{length} \\ \text{weight} \end{bmatrix}$$

Training sets have been used, and statistics developed as follows:

$$\boldsymbol{\mu}_{\text{gasket}} = \begin{bmatrix} 2.8 \\ 4.0 \end{bmatrix} \quad \boldsymbol{\mu}_{\text{flange}} = \begin{bmatrix} 2.1 \\ 3.0 \end{bmatrix}$$

$$C_{\text{gasket}} = \begin{bmatrix} 0.01 & 0 \\ 0 & 0.01 \end{bmatrix} \quad C_{\text{flange}} = \begin{bmatrix} 0.02 & 0.03 \\ 0.03 & 0.06 \end{bmatrix}$$

Apply Eq. 13.7 and distinguish whether the measurement

$$\mathbf{x} = \begin{bmatrix} 2.2 \\ 3.8 \end{bmatrix}$$

is most likely a flange or a gasket.

# 14

---

# *COMPUTATIONAL ARCHITECTURES*

In the previous chapters, we have described a number of algorithms for controlling various aspects of the motion of a robot. Many of these calculations have real-time constraints in that they must be completed within some small time period (e.g., 15 milliseconds) or the robot control loop may become unstable. In addition, these calculations are often complex and involve evaluation of trig functions, multiplication and/or inversion of matrices, or processing of large quantities of data such as a picture.

Nowhere, prior to this chapter, have we been concerned with how so many complex calculations may be performed in so short a time. In this chapter, we will suggest some ways this may be accomplished. The suggestions include various ways in which to organize the hardware to meet this objective and also some recommendations that enable software to provide surprisingly fast results.

## 14.1 SOFTWARE SPEED-UPS

In this section, we will discuss techniques for increasing the speed with which computations may be performed in a conventional computer. These techniques will take advantage of the unique properties of many of the calculations which occur in robotics.

### 14.1.1 Using Integer Arithmetic

In this section we will represent numbers as integers plus fractions and avoid the use of floating point ("real") representations for numbers.

We will accomplish this by a technique known as *scaling*, in which a mixed number (e.g., 22.75) is represented by an integer (e.g., 2275) which is multiplied, or scaled, by a constant. We will discover that if the scaling is constant, considerable speed advantage can be gained over floating point. Even with the advent of high-speed floating point processors, integer arithmetic is typically 5 to 10 times as fast as floating point.

Furthermore, even while using purely integer arithmetic, fractional values can still be handled. To see this, consider an example of a problem in which we know that numbers can never exceed 255 and can never be negative. (Other cases are also easily handled by similar approaches.) We will use a 16-bit integer to represent the numbers and then *interpret* such a number as 8 bits of magnitude (the left byte) and 8 bits of fraction (the right byte). All the integer arithmetic instructions operate correctly using this interpretation, as Example 14.1 shows. Higher accuracy is easily achieved by using, for example, 32-bit integers.

**Example 14.1  Multiplication of Fractions**

Using a 16-bit, unsigned integer representation for numbers, with the binary point between the bytes, show the effect of multiplying $9.5 \times 2.5$.

**Solution:**   In this representation, we write 9.5 as $9 + \frac{1}{2}$, so that we may use a binary rather than a decimal representation. The 16-bit form of 9.5 is then

$$00001001.10000000$$

The binary point is explicitly shown for clarity. Ones to the *right* of the point represent $\frac{1}{2}^k$ where $k$ is the number of places to the right of the point. Similarly, 2.5 is

$$00000010.10000000$$

Multiplying these two 16-bit numbers yields the 32-bit result:

$$0000000000010111.1100000000000000$$

We observe that the binary point has moved to the sixteenth bit position. This is because our representation is really a scaling of a fractional quantity into an integer value by multiplying by 256. Since both multiplier and multiplicand are scaled by 256 (8 bits) the product is scaled by 16 bits. To find the answer in our 16-bit representation, we extract the middle 2 bytes

$$00010111.11000000$$

The answer may then be interpreted as the number left of the binary point $10111 = 23$ plus the binary fraction to the right, $\frac{1}{2} + \frac{1}{4} = 0.75$. And, as expected, $9.5 \times 2.5 = 23.75$.

As Example 14.1 shows, integer arithmetic can be used to gain speed and, by proper scaling, can still represent fractional values. Before we can use such techniques, we need to know a priori that the loss of accuracy inherent in scaling will not seriously impact the performance of the system; that is, we need to have some idea of the size range of the numbers we will get, not only as final results, but as intermediate results as well. For example, when inverting a matrix, intermediate results may vary significantly in magnitude from the elements of the original matrix. In inverting a transform matrix, we have no problems, since Eq. 6.23 provides a closed-form solution for the inverse. However, when an inverse must be computed numerically,* as was discussed in Section 8.3.1 with regard to the Jacobian, one must be very careful in using integer arithmetic to avoid loss of precision.

### 14.1.2 Computing Trig Functions

In commercial systems, trigonometric functions are evaluated by several techniques including rational functions or Chebychev polynomials. The arithmetic involved is usually done in floating point, resulting in high accuracy and low speed.

In robotic applications, however, the most appropriate mechanism for computing trig functions is by table lookup. For example, the values of $\sin(i)$ are simply stored in a table for $i = 0 \ldots 89$ degrees. This table of 90 words provides instantly the value of the sine of an angle to a $1°$ precision. In fact, the 90 words required is probably less memory than that required by a program to compute the function explicitly.

If higher accuracy than the nearest degree is required, two options exist. The first is linear interpolation between table entries. Sine and cosine are quite linear between $1°$ intervals, and interpolation gives excellent results. Tangent is also well behaved away from $90°$.

Another interpolation technique makes use of the trigonometric identity for the sine for the sum of two angles. Specifically, we make use of the fact that

$$\sin(\alpha + \beta) = \sin \alpha \cos \beta + \cos \alpha \sin \beta \qquad (14.1)$$

We segment the angle into two parts, an integer, $\alpha$, and a fraction, $\beta$, and apply Eq. 14.1 If we use a sixteen-bit representation for angles, scaled so as to have eight-bit integer and fractional parts, both parts may be stored in lookup tables containing only 256 entries each. Exact evaluation of Eq. 14.1 then requires four table lookups, two multiplies, and one add. An approximate evaluation can be obtained even faster; the angle is scaled so that the fractional part is a small angle, and for small

---

*Recall, as we discussed in Section 8.3.1, that we almost never explicitly compute an inverse numerically; rather, we solve a system of linear equations, which requires less computation than a full inverse.

$\beta$, we have

$$\cos \beta = 1 \qquad (14.2)$$

and

$$\sin \beta = \beta \qquad (14.3)$$

This is illustrated numerically in the following example.

### Example 14.2 Interpolating Trig Functions

Assume that you have tabulated the values of the sine function in 0.01 rad (radian) increments. (The switch from degrees to radians is essential here, as will become clear.) That is, you know the value of sin (0.80 rad) and sin (0.81 rad). Show how to use Eq. 14.1 to find sin (0.8034 rad).

**Solution:**    First, write

$$0.8034 \text{ rad} = 0.8 \text{ rad} + 0.0034 \text{ rad}$$

We know that

$$\sin (\alpha + \beta) = \sin \alpha \cos \beta + \cos \alpha \sin \beta$$

Therefore,

$$\sin (0.8034) = \sin (0.80) \cos (0.0034) + \cos (0.80) \sin (0.0034)$$

For such small angles, cos (0.0034) = 1 and sin (0.0034) = 0.0034, and we have sin (0.8034) = sin (0.80) + 0.0034 cos (0.80). This interpolation technique requires two table lookups, one multiply and one add. Conventional linear interpolation requires two lookups, four adds, one divide, and one multiply if the slopes are not stored in the table. If the slopes are stored, linear interpolation requires two lookups, one multiply and one add. In fact, since the slope of the sine function is the cosine, this technique is exactly equivalent to linear interpolation with stored slopes.

Surprisingly, in actual robot system implementation, the calculation of trig functions is one of the fastest of the operations.

### 14.1.3 Matrix Operations

The literature (Stewart, 1973) is quite complete with respect to efficient matrix operations. We will make only a few observations regarding the specific data structures described in this book.

Heuristics are easily developed to make use of the fact that the fourth row of a homogeneous transform is a constant.

Matrix inversion is almost always more efficient if it can be solved in closed form rather than numerically evaluated. We have followed this philosophy both in Chapter 7 with the inverse kinematic transform and in Chapter 8 with the inverse Jacobian. Although the closed-form solution may appear very complex, most of the terms also occur in the forward transform and, therefore, are already calculated and stored. For some arms, a closed-form solution, particularly of the Jacobian, may be impossible. In this case, the inversion must be performed numerically, or simple control techniques, not requiring the inverse Jacobian, must be used. Having made these statements we remind the reader of the analysis performed in Chapter 8 in which we demonstrated that the computational complexity of inverting a $6 \times 6$ matrix, while somewhat more than that of a closed form solution, is not necessarily so bad as to rule out numerical techniques completely.

Throughout this book, we have consistently used the homogeneous transform as the mechanism for representing arm configurations. This is because the homogeneous transform provides the best mechanism we have found for representing the fundamental concepts of robotics so that they may be easily grasped by students. As was mentioned in Chapter 6, other formalisms exist for representing arm configurations, some of which may be more efficient from a computational point of view. In particular, Taylor (1983) makes a strong case for the use of quaternions.

In summary, we may state that many of the techniques described in the numerical analysis literature have been developed to produce high accuracy and stability or robustness. In robotics, speed is of the essence, whereas accuracy, although important, is not as difficult to maintain. Thus, we may substitute simple, fast methods over complexity and high precision in many cases.

## 14.2 COMPUTATIONAL ALTERNATIVES: HARDWARE

*Distributed computing,* the use of several computing units, interacting in real time to provide the required computational speed and power, finds an immediate application in robotics. The literature on distributed computing is diverse (Anderson and Jensen, 1975), and many, if not all, of the concepts described are directly applicable to robotics. In this chapter, we will mention some of the options generically and provide a couple of examples. The serious robot system designer should refer to the cited references for more details.

### 14.2.1 The General-Purpose Computer

Throughout this book, we have referred to computation as if it were being performed by a single general-purpose computer. Let us consider the feasibility of this by taking a closer look at two simple functions: reading encoders for position information and computing joint servo error signals.

### Reading Encoders

The rate at which data comes from an encoder depends upon the rotational speed of the joint, the resolution of the encoder, and the algorithm used for analyzing the signals. If we assume quadrature decoding (a new count for each of the four states of an incremental encoder, see Chapter 2), a 2500-count/rev encoder, and a 1-rev/sec maximum joint speed, we find a new piece of data every 100 $\mu$sec. Using interrupts, the absolute minimum set of instructions would be

REGISTER SAVE
BIT TEST
JUMP
INCREMENT (OR DECREMENT)
REGISTER RESTORE
RETURN

It is hard to imagine this process requiring less than 5 to 10 $\mu$ sec on any computer. As a reasonable estimate, say, 5 $\mu$s. With six joints to control, we find that at maximum speed, just keeping track of position requires $5 \times 6/100 = 30$ percent of the computer resources.

### Computing Servo Error Signals

With a PD (proportional derivative) controller, we must compute

$$T_\theta = K_e(\theta_d - \theta) - K_d\dot{\theta}$$

The program to implement this calculation is

LOAD $\theta_d$
LOAD $\theta$
SUBTRACT
MULTIPLY $(\theta_d - \theta)$ BY $K_e$
LOAD $\dot{\theta}$
MULTIPLY $\dot{\theta}$ BY $K_d$
SUBTRACT
OUTPUT
RETURN

Optimistically, this program would take 15 $\mu$s. For stability (Paul, 1982), the joint servo needs to be calculated roughly every 2ms.* Thus,

---

*This number varies significantly with the structural resonances of the manipulator. Paul (1982) actually determined a figure of 3.3 ms for the Stanford manipulator.

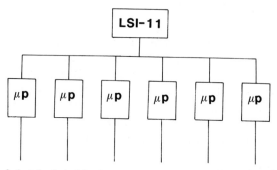

Joint 1  Joint 2  Joint 3  Joint 4  Joint 5  Joint 6

Figure 14.1   PUMA architecture.

joint servo calculation takes on the order of

$$\frac{6 \times 15 \times 10^{-6}}{2 \times 10^{-3}} = 45 \times 10^{-3} = 4.5 \text{ percent of computer resources}$$

These estimates vary tremendously with implementational details (see Problem 2). The point, however, is that a significant amount of the computing involved in robotics is simple, localized, and easily distributed among processing elements.

### 14.2.2 Distribution of Computing by Joints

The Unimate PUMA architecture exemplifies the concept of distribution by joints. This architecture is shown in Figure 14.1. A separate processor is attached to the input and output transducers of each joint. This processor maintains current angular position and velocity for that joint and servos that joint toward a set point provided by the central processor. The central processor reads angular position from the joint processors, computes kinematic transformations, and provides set points to the joint processors.

By performing all the kinematics in the same processor, this architecture provides for simplicity of design and programming, while offloading concurrent control of joints to parallel processors.

### 14.2.3 Distribution by Function

The system shown in Figure 14.2 (Evans, Snyder, and Gruver, 1978) exemplifies the architectural philosophy of distribution of computation by function. Like the PUMA, this system associates one processor with each joint. In this system, however, the calculation of kinematics is also distributed. Each processor maintains the position

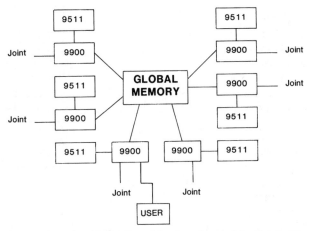

**Figure 14.2**    Robot control distributed by function and by joint. The system which was finally implemented had only four processors rather than the six shown in this figure.

and velocity of its joint, as well as the various trig functions of those parameters. These quantities are stored in a multiported memory (Evans, 1979) and hence are immediately available to the other processors. Calculation of the other kinematic terms is distributed uniformly among the processors.

This scheme distributes the computational burden more uniformly, at the expense of a complex piece of hardware (the multiported memory) and more complex software. An immediate extension to and improvement of this architecture would be as shown in Figure 14.3, disassociating the servo calculations from the kinematics. Figure 14.3 is, in fact, just the PUMA architecture with the single central processor replaced by an array of processors and common memory.

### 14.2.4  Custom Hardware

As of this writing, we are witnessing the rapid development of LSI and VLSI devices which, when suitably combined with general-purpose computing, can provide tremendous increases in performance. For example, the AMD9511, which has been available for several years, provides full 32-bit floating point capability and is designed for easy interfacing to 8-bit microprocessors. The Intel 8087 provides high-speed arithmetic and direct coupling to the 8086 bus. TRW and AMD both advertise multiplier chips that perform 16-bit integer multiplies in just a few nanoseconds.

The state of the art is changing so rapidly that coordinate transform chips can be anticipated in the very near future, and a chip which

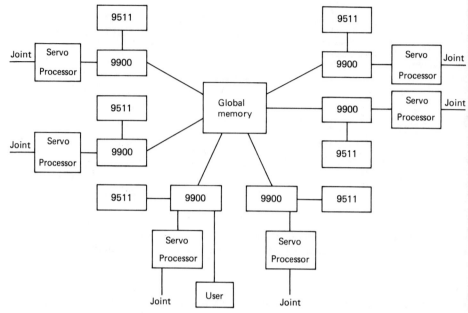

**Figure 14.3**   A proposed addition to the architecture of Figure 14.2 to decouple the servo calculations from the distributed computation of kinematics.

computes the Jacobian and even the inverse Jacobian in real time is a real possibility. We can anticipate that such chips will have the following characteristics:

1. Fast, repetitive calculations (typically those which do not require conditional branches)
2. Easy interface to a general-purpose processor's bus system
3. Direct control by such a processor

## 14.3 SEMICUSTOM HARDWARE

*Microprogramming* does not refer to the programming of microprocessors. Instead, the term refers to the sequencing of the most primitive operations of a digital system such as a computer. For example, the instruction

ADD  A,B

which adds a register named A to a register named B and stores the result in A could be broken down into the following, primitive micro-operations:

1. Enable output of register A onto bus X.

**2.** Enable output of register B onto bus Y. (We assume buses X and Y are inputs to an arithmetic-logic unit and bus Z is the output of the ALU.)

**3.** Signal ALU to add.

**4.** Latch the number on bus Z into register A.

A single microoperation may be represented by a single bit in a *microinstruction*. Microinstructions are typically wide; 60 bits is not unusual. Therefore, a single microinstruction causes many things to occur simultaneously.

Microinstructions are stored in a high-speed ROM. This ROM then controls the operation of the computer. A sequence of microinstructions is called a microprogram. It is, in fact, the microprogram that defines how the computer executes instructions, as we saw with the ADD instruction. By supplementing or changing the microinstruction, instructions may be created or redefined.

By modifying the address to the ROM, the microprogram may contain jumps. The design of a system for sequencing through a microprogram is complex but is made somewhat easier by the availability of *microsequencer chips*. Families of such chips are available, including the ALU. The AMD 2900 series (Mick and Brick, 1980) is probably the best known.

Microprogramming can result in extremely fast and efficient code. It is also extremely difficult to program. The development cost of a microprogrammed controller is typically higher than development cost for a software controller, but less than the costs for a hard-wired controller. Of course, because the control sequence is stored in a ROM, the microprogrammed controller is much easier to modify and enhance than is the hard-wired controller.

Some commercial robots achieve the computational speed required for conveyor tracking by using an AMD 2900 series of microprogrammed chips.

## 14.4 HIERARCHICAL CONTROL

The term *hierarchical control* is perhaps best typified by the command structure of the military, in which the generals pass orders to the officers, who pass orders to the sergeants, who pass orders to the troops. In such a hierarchy, control is supposed to flow down from the top, and reporting is intended to flow up from the bottom.

A hierarchical decomposition of a control task, particularly in robotics, has the same flavor. We have already seen one example of a hierarchy, in the control of the PUMA robot. In that system, a central processor computed the kinematics and provided set points to the joint servos. A number of robot system designers have approached their tasks in similar ways.

Although most authors would agree on the basic philosophy of hierarchical control, there is considerable difference in methods used to implement hierarchy. For example (Mesarovic, Macko, Takahara, 1970), one could consider the control of a multivariate system in a rigorous mathematical form and attempt to find optimal, or suboptimal, techniques for controlling the entire system. If such a technique were properly structured, it could be implemented in a hierarchically struc-tured array of processors. Such mathematical decompositions are the topic of considerable current research in control and optimization theory.

In Barbera, Albus, and Fitzgerald (1979) and Albus (1977), the con-cepts of hierarchical control are elucidated most clearly as they apply to current practice in robot control, and we restate those concepts here.*

An advanced robot system may be partitioned into three distinct modules, as shown in Figure 14.4. The program module accepts inputs from the user, either interactively or in the form of a robot program (see Chapter 15). This task description is translated into a set of lower-level commands, perhaps including conditional branches, much as a compiler might convert from FORTRAN to machine language.

The location module serves to relate symbolic names of locations, for example, "clamp," to Cartesian descriptions of points in the work space.

The control system module is responsible for controlling the mo-tion of the robot so that it can achieve the tasks described by the pro-gram module, at the locations described by the location module. The control system module has a real-time requirement and must be further decomposed.

### Real-Time Control

Figure 14.5 represents a hierarchical decomposition of the real-time control functions. The precise number of levels in this decomposition, as well as the exact responsibilities of each level, are details that are the concern of the system designer and are unimportant at this point. There are, however, some important observations to be made with regard to the general structure of Figure 14.5.

First, at the higher levels, the control functions are independent of the arm geometry. That is, determination of the path (Chapter 9) is

*See Saridis (1983) for an alternative approach.

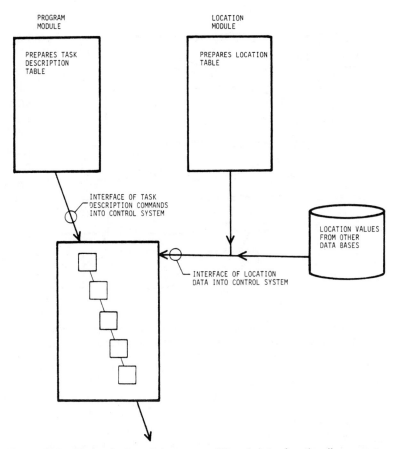

**Figure 14.4**   (Barbera)   Overall   system   partitioned   into   functionally   separate groups.  Two major communication interfaces are defined.  The program module to control system interface is in the form of a table of task description commands that form the input to the highest level of the control system.  The location data-to-control system interface is in the form of a table of location point values specified in an arbitrary Cartesian coordinate system.  This table can be provided either by the location module or by other data bases such as CAD or NC PART CLDATA file.  (*U.S. Bur. of Standards*)

performed in Cartesian coordinates, relative to the workspace frame. Locations in this frame are not necessarily constant, for it may be necessary to specify locations in a moving frame such as on a conveyor (which means that the location module must be capable of real-time updates); however, at these higher levels, task description does not require knowledge of the robot geometry.  For this reason, the same algorithms and software may be used for many different robots, rather like FORTRAN may be run on both a TRS-80 and a CRAY.

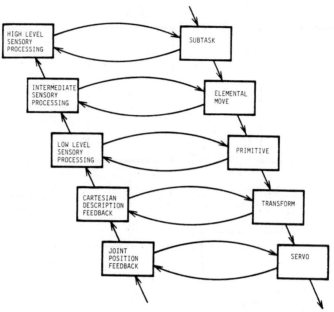

**Figure 14.5** (Barbera) Two parallel hierarchies: one, a sensory processing hierarchy that performs the data reduction of the sensory information into a form usable by the control algorithms; the other, a control hierarchy that provides a real-time update of its outputs based on the feedback data from the sensory processing hierarchy. (*U.S. Bur. of Standards*)

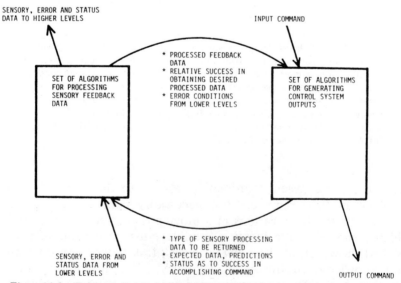

**Figure 14.6** (Barbera) Detailed view of the types of interactions between the sensory and control hierarchies at one level. (*U.S. Bur. of Standards*)

At the lower levels of the control hierarchy, the details of the algorithms become closely related to the arm geometry, once the kinematic transformations have been performed, and the control algorithms may vary from one robot to another. We see that happening at the two lower levels of Figure 14.5. The lowest level compares set points (provided by the next level) with joint positions (provided by sensors) and moves the joint to reduce the error. The next level computes the kinematic transformations and provides those set points to the bottom level.

We also observe, from Figure 14.5, that there are really two hierarchies involved: a top-down hierarchy of control and a bottom-up hierarchy of status. Figure 14.6 shows this most clearly by detailing a single level in the structure.

Since each level in the hierarchy is well defined functionally, the robot control task can be divided into processing modules in which each module has well-defined inputs and outputs and, therefore, can be associated with an independent processing element.

## 14.5 CONCLUSION

We began this chapter by observing that the fundamental difficulty in robotics is not the complexity of the mathematics (with the possible exception of optimal control including kinetics) but, rather, the speed with which solutions must be computed.

We first addressed the issues associated with using a general-purpose computer and increasing its speed by software techniques until that speed met the requirements of robotics. Such software speed-ups are well known in the literature of numerical methods.

We then investigated special-purpose hardware and microprogramming and concluded that this area is evolving so rapidly that many high-speed, special-purpose chips can be expected in the very near future.

Finally, we described a philosophy of distribution of both control and feedback in a robotic system, a philosophy that can guide the system designer in his or her initial decision making.

## 14.6 SYNOPSIS

### Vocabulary

You should know the definition and application of the following terms:

binary point
hierarchy

Gaussian elimination
microsequence
microprogram


## 14.7 REFERENCES

ALBUS, J. S. Control concepts for industrial robots in an automated factory. Society of Manufacturing Engineers Technical Paper MS77-745, 1977.

ANDERSON, G. A. and JENSEN, E. D. Computer interconnection structures: Taxonomy, characteristics, and examples. *Computing Surveys, 7* (4), December 1975.

BARBERA, A. J., ALBUS, J. S., and FITZGERALD, M. L. Hierarchical control of robots using microcomputers. 9th International Symposium on Industrial Robots, 1979.

EVANS, P. F. A Microprocessor-Based Distributed Processing System. MS Thesis, North Carolina State University, 1979.

——, SNYDER, W., and GRUVER, W. A high-speed multiple microprocessor architecture for real-time control applications (TR 78-17). Computer Studies Department, North Carolina State University, 1978.

MESAROVIC, M. D., MACKO, D., and TAKAHARA, Y. *Theory of hierarchical multilevel control.* New York: Academic Press, 1970.

MICK, J., and BRICK, J. *Bit-slice microprocessor design.* New York: McGraw-Hill, 1980.

PAUL, R. P. *Robot manipulators.* Cambridge, Mass.: M.I.T. Press, 1982.

SARIDIS, G. N. Intelligent robotic control. *IEEE Transactions on Automatic Control, 28*, pp. 547–556, May 1983.

STEWART, G. W. *Introduction to matrix computations.* New York: Academic Press, 1973.

TAYLOR, R. H. Planning and execution of straight line manipulation trajectories. *IBM Journal of Research and Development, 23*, pp. 424–436, 1979. Also in *Robot motion, planning and control,* Brady, M., Hollerbach, J., Johnson, T., Lozano-Pérez, T., and Mason, M., eds. Cambridge, Mass.: M.I.T. Press, 1983.


## 14.8 PROBLEMS

1. (a) Show the binary representation for 14.75, using the 16-bit form demonstrated in Example 14.1. (b) Show the binary representation for 12.5, using the same representation. (c) Using binary, integer arithmetic, multiply the two previous results and demonstrate that the answer is correct.

2. (a) The following is an interrupt service routine for an MC68701 processor. Evaluate the amount of time required for this routine to run. At the discretion of your instructor, you may use either the true instruction execution times, or you may assume that each instruction requires 1 μs. (b) Supposing that an in-

terrupt occurs every 200 $\mu$s, determine the percentage of computer time utilized in servicing interrupts.

```
NMI   LDAA   PIDATA
      ANDA   #$03
      CMPA   #$01
      BEQ    CU
      CMPA   #02
      BEQ    CD
      RTI
CD    LDX    SUM
      DEX
      STX    SUM
      RTI
CU    LDX    SUM
      INX
      STX    SUM
      RTI
```

3. Determine the number of arithmetic operations required to invert a transform matrix, using the method described in Chapter 6.

4. Suppose that the architecture of Figure 14.2 is used but is simplified to control the $\theta$-$r$ manipulator. That is, three processors communicate via a common memory. Two of the processors are associated with the joints, and the third is reporting Cartesian position of the hand to an outside data receiver. Each joint processor measures the position and velocity of its joint. Make reasonable assumptions, state those assumptions, and determine the number of accesses to the global memory that must be made every second. Assume the Cartesian position must be updated every 20 ms.

# 15

---

# ROBOT
# PROGRAMMING
# LANGUAGES

Robot programming languages* are almost as varied as the robots they are designed to manipulate. The language emphasis varies from simplistic point-to-point motion to complex task-oriented problem solving. The languages available today can be divided into five loosely formulated levels. Overlaps between levels occur but do not interfere with the basic features of each. Figure 15.1 contains a breakdown of 15 languages into these five language levels.

In this chapter, the five levels are described. Then, an example task, unloading a conveyer, is programmed in each language.

## 15.1 THE HARDWARE LEVEL

The hardware level represents the lowest level of the robot language hierarchy. The commands at this level are highly dependent on the physical structure of the robot. The emphasis is on converting joint error signals to torques and forces. The control of each joint is handled independently by a microcomputer, some hard-wired device, or by a task in a multitasking system. The languages in use at this level are the traditional computer languages such as C or assembly language, and they are used to implement real-time control functions.

*The author is deeply indebted to Susan Bonner (Bonner, 1983) for her help and cooperation in the preparation of this chapter. Much of the material contained herein is extracted from her thesis.

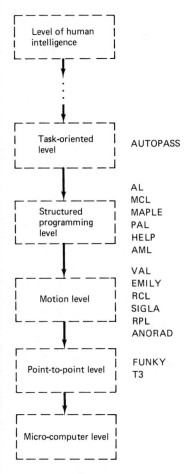

Figure 15.1    (Bonner) Language levels.

## 15.2 THE POINT-TO-POINT LEVEL

The most common type of robot programming language available on the market today is the point-to-point class. Programmed robot control is provided by saving a series of points obtained by guiding the robot through its motions, either with a teach pendant or by moving the robot manually through its motions, as is done in spray painting robots (see Chapter 9). Most systems also include specialized buttons that allow editing of stored programs or interaction with external signals. Both the T3 language and IBM's FUNKY are classified as higher-level guiding systems.

T3 is operated manually by pressing buttons on the teach pendant or entering specialized commands on the console. The robot may be controlled through the use of Cartesian, cylindrical, or joint coordinate systems. Once stored, a program may be edited by stepping forward or backward through the program and using insert or delete command functions. The program interacts with external signals simply by waiting for the external signal to occur. Instead of a teach pendant, a FUNKY system has a joystick that controls the motion of the robot. Control of the system is similar to a tape recorder with PLAY, ERASE, RECORD, REVERSE, and FAST FORWARD modes. As with T3, programs can be edited by stepping through the program to the appropriate point and inserting or deleting commands. FUNKY does provide somewhat more capability in sensing than T3, including, in particular, a command to center the gripper about an object, using a touch sensor mounted on the hand.

**Example 15.1  Palletizing with FUNKY (Grossman, 1977)**

Figure 15.2 illustrates a robot operation. FUNKY does not provide commenting capabilities, nor does it provide branching. It creates a file of commands in a subset of the motion-level language EMILY. In a system with EMILY support, the file could then be edited to include conditional branches or sensing of external signals.

In Figure 15.3, parameters enclosed in angle brackets represent robot joint positions. These positions are determined interactively by locating the robot in the desired configuration and pressing the *define* button. The commands themselves are entered by pressing buttons on the command keyboard.

The ABSOLUTE command specifies motion to an absolute position, determined, as above, by pressing the "define" button. Using the STORE command allows that $N$-tuple of joint positions to be stored under a symbolic position num-

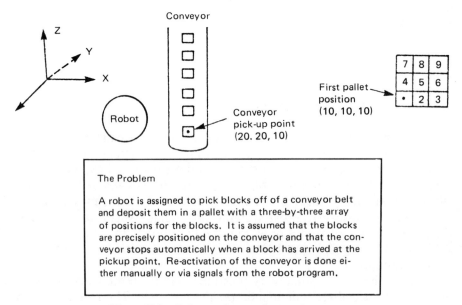

Figure 15-2   Robot work space for sample test case.

FUNKY:

```
        ABSOLUTE   <starting-point>
        STORE 1
        ABSOLUTE   <conveyor-approach-point>
        STORE 2
        ABSOLUTE   <conveyor-pick-up-point>
        STORE 3
        CENTER 8
        RECALL 2
        ABSOLUTE   <first-pallet-position>
        STORE 4
        ABSOLUTE   <first-drop-position>
        RELEASE 8
        RECALL 4
        HALT   0
        RECALL 2
        RECALL 3
        CENTER 8
        RECALL 2
        ABSOLUTE   <second-pallet-position>
        STORE 4
        ABSOLUTE   <second-drop-position>
        RELEASE 8
        RECALL 4
        HALT   0
        RECALL 2
            .
            .
            .

        HALT   0
        RECALL 2
        RECALL 3
        CENTER 8
        RECALL 2
        ABSOLUTE   <ninth-pallet-position>
        STORE 4
        ABSOLUTE   <ninth-pallet-position>
        RELEASE 8
        RECALL 4
        HALT   1
```

**Figure 15.3**   (Bonner) Palletizing with FUNKY.

ber. The STORE command actually pushes the location of the previous command onto one of nine available stacks. The RECALL command causes motion to the position at the top of the indicated stack. The CENTER and RELEASE commands position and activate the gripper. In this example, the HALT commands are used in lieu of conveyor start-stop commands.

## 15.3 THE MOTION LEVEL

The motion level of robot programming provides a language in which to implement the simple point-to-point control strategies described in the previous section. When compared with the capabilities of the point-to-point level, the motion-level languages exhibit the following characteristics:

1. Simple branching has been added.
2. Subroutines (generally with parameter passing) can be used.
3. Sensing capabilities are more powerful.
4. The capability to initiate parallel activities is introduced.
5. The capability of representing and manipulating frame descriptions is included.

Interpreters are by far the most popular mechanism for implementing robot languages. This is because the programmer may enter a single command and call for immediate execution of that command. In this way he or she may debug the program a step at a time.

ANORAD, RCL, SIGLA, EMILY, VAL, and RPL are classified here as motion languages. All these languages are based on interpreters or assemblers except RPL, which has a compiler. They all provide simple conditional branching. EMILY, VAL, and RPL support DO loops. Motion can be specified either in joint coordinates directly or in Cartesian coordinates. Details of how Cartesian coordinates are specified vary from one system to another. All the languages provide motion to absolute positions. ANORAD, RPL, and VAL also provide relative and straight-line motions.

In addition to subroutines, most languages at this level also provide some type of INCLUDE function. The INCLUDE statement allows the user to specify a file of statements to be read in at the position of the INCLUDE statement and treated as if that file were part of the program. The INCLUDE may be accomplished by allowing a file to be run as a subroutine or by literally reading it as part of the source. Again, the fact that these programming languages are implemented as interpreters provides convenience.

The motion-level languages vary widely in their sensing capabilities. ANORAD, RPL, SIGLA, and VAL are provided with simple, binary touch sensing commands. EMILY supports several different kinds of tactile sensors, including touch sensors in the hand, a "whisker" on one finger, an ultrasonic proximity detector, and an infrared proximity detector. All the sensors are treated as on-off conditions by software. Unlike EMILY, RPL supports only very simple touch functions, but it includes support for a relatively sophisticated vision capability.

Parallel processing at this level exists but is not very sophisticated. For example, both SIGLA and EMILY allow simultaneous execution of different program files on different robots. EMILY provides an explicit SYNCH command to converge two concurrent processes. SIGLA provides some collision avoidance by allowing the user to specify the work volume for each arm. Real-time collision avoidance [see, for example, the work of Lozano-Pérez (1980)] is not supported at this level.

Some of the motion-level languages provide support for symbolic representation and manipulation of frames. This will be discussed in more detail in the next section.

**Example 15.2.  Palletizing with VAL (Unimation, 1979)**

The VAL program presented in Figure 15.4 illustrates the same palletizing function as that illustrated in Figure 15.2.

Lines 1 and 2 initialize integer variables which are used to keep track of how many parts have been loaded onto the pallet in both the $x$ and $y$ directions. The GOSUB on line 3 functions just as its equivalent in BASIC, to call a subroutine which will unload one part from the conveyor and place that part on the pallet. The command SHIFT PALLET does not cause any motion of the physical pallet; rather, it redefines a coordinate frame named PALLET. Thus, the next time the robot moves to the position named PALLET, it will be a new position. The SHIFT command on line 5 implements a translation of $x = 100.0$ mm, $y = 0$, and $z = 0$.

Inside the subroutine, the robot first approaches the frame named CON. The APPRO function causes motion to a point displaced (in this example) 50 mm out the $z$ axis of the named frame and oriented such that the $z$ axis of the hand is aligned with the $z$ axis of the named frame. In addition to specifying the desired position and orientation, the APPRO command specifies joint interpolated control.

The command on line 13 allows the robot to wait for a signal from an outside source, in this case, a limit switch, indicating that the conveyor is ready. The manipulator then moves (MOVES indicates Cartesian straight-line motion) to the grasp position and closes the gripper. If the hand closes to less than the minimum anticipated distance (25 mm), an error is assumed and the program is halted.

The DEPART command specifies Cartesian motion along the $z$ axis of the hand frame to the departure point, in this case 50 mm out. The command on line 17 could equally well be APPRO PALLET, 50; however, the alternative form is used to demonstrate the capabilities of manipulating frames in VAL. MOVE specifies

```
1.              SETI      PX = 1
2.              SETI      PY = 1
3.   10         GOSUB     100
4.              IF PX = 3 THEN 20
5.              SHIFT PALLET BY 100.0,0,0
6.              GOTO 10
7.   20         IF PY = 3 THEN 40
8.              SETI PX = 1
9.              SETI PY = PY + 1
10.             SHIFT PALLET BY -900.0, 100.0,0
11.             GOTO 10
12.  100        APPRO CON,50
13.             WAIT CONRDY
14.             MOVES CON
15.             GRASP 25
16.             DEPART 50
17.             MOVE PALLET:APP
18.             MOVES PALLET
19.             OPENI
20.             DEPART 50
21.             SIGNAL GOCON
22.             SETI PX = PX + 1
23.             RETURN
24.  40         STOP
```

**Figure 15.4**    Palletizing with VAL.

joint interpolated motion (as distinguished from MOVES) to a point defined by its argument. In this case the colon indicates that the argument of MOVE is a transform which is the product of two other transforms, the current pallet frame, and a transform specifying a displacement of 50 mm out the $z$ axis. In this example, APP would be the constant transform:

$$APP = \begin{matrix} 1 & 0 & 0 & 0 \\ 0 & 1 & 0 & 0 \\ 0 & 0 & 1 & 50 \\ 0 & 0 & 0 & 1 \end{matrix}$$

Finally, a VAL program can output signals to external devices, in the case of line 21, a command to start conveyor motion.

## 15.4 STRUCTURED PROGRAMMING LEVEL

The structured programming level incorporates structured control constructs into the robot language and provides extensive use of coordinate transformations and frames. Other characteristics of these languages include complex data structures, improvements in sensor processing, parallel processing capabilities, and the extensive use of predefined state variables.

In this discussion, HELP (General Electric, 1982), PAL (Takase, 1981), MCL (McDonnell Douglas, 1980; Baumann, 1981), MAPLE (Darringer and Blasgen, 1975) and AL (Sahid, 1980) are included in this level of languages. Although PAL is not truly a structured language, it is included because it provides some structured control constructs and uses coordinate transformation to a great extent. HELP does not make use of transforms at all, but because it has structured programming constructs, it is included in this level.

Most of the languages at this level are based on existing structured languages such as ALGOL and PL/1. These structured base languages provide the robot languages with many powerful control aids and data structures, including constructs such as "if . . . then . . . else" and "while . . . do."

In languages at both the structured and motion level, frames may be represented by a single variable. For example, in PAL (Takase, 1981),

MOV BRH + A

represents a motion to a frame defined symbolically as BRH and mnemonically as *bracket hole*. A is an arbitrary offset point, typically

10 cm up the $z$ axis.  Thus, A represents the matrix

$$
\begin{array}{cccc}
1 & 0 & 0 & 0 \\
0 & 1 & 0 & 0 \\
0 & 0 & 1 & 10 \\
0 & 0 & 0 & 1
\end{array}
$$

and the + sign represents matrix multiplication.  Thus the expression BRH + A represents an approach point defined 10 cm up the $z$ axis of the bracket hole.

In VAL, one might write

MOVES BRH: A

where the : operator means *relative to* and implies matrix multiplication of frames.

In addition to structured programming constructs, the typical language at this level has definable subroutines with parameter passing. They all, with the exception of HELP, are provided with complex data structures:  PAL has transforms as its basic element; MAPLE allows the definition of points, lines, planes, and frames; MCL provides for points, vectors, and frames; and AL utilizes scalars, vectors, rotations, frames, and transformations.  MAPLE and MCL break transformations into positional and rotational parts.  MAPLE uses MOVE BY and ROTATE BY to perform transformations.  MCL defines the positional part of a transformation as a point and uses two vectors to provide the orientation.

MCL and AL also provide means of fixing frames together so that transforms applied to one part are automatically applied to another. When the relationship between two parts is a constant due to the fact that they are fastened together, this feature provides that if the second frame moves, the first is automatically moved also.

While sensor commands at the structured level are similar to those at the motion level, there is a wide variation in the sophistication of the sensing capabilities among the different languages at the structured level.

Parallel processing is expanded at the structured level.  All have semaphore primitives that are activated only when a given event occurs. MAPLE has an IN-PARALLEL construct for use when the order of execution of commands is irrelevant.  AL has the COBEGIN and COEND constructs for synchronization between two arms.  In all these cases, the user is responsible for collision avoidance.

Motion on the structured level is generally defined in terms of transformations on the robot hand frame. In PAL and AL, motion is specifically directed in terms of the transform. PAL uses mathematical symbols of + and – and the MOV mnemonic to cause motion, whereas AL uses the MOVE TO statement. AL motion can be made more specific through the use of clauses which define intermediate points, approach vectors, velocities, and durations. In MCL and MAPLE, motion is separated into rotations and translations. The MCL GOTO statement will perform a translation and rotation, but both must be specified (as a point and two vectors). MAPLE provides MOVE BY for translations and ROTATE BY for rotations. HELP is an exception because it provides motion in terms of the more primitive joint coordinates.

**Example 15.3. Palletizing with PAL**

The routine in Figure 15.5 uses mathematical notation and transformation to perform the pallet task. Conveyor motion is assumed to take care of itself. The routine begins the task by initializing all needed transforms. $Z$ is the base transform. $t6$ represents the robot configuration and is therefore known. $e$ is a transformation representing the hand with respect to the end effector. *con* and *pal* are conveyor and pallet position, respectively. *app*, *pos*, and *row* are transforms to be applied to the frames to create an approach position, move to a new pallet position, and move to a new pallet row. The *tol* and *arm* variables cause motion to the indicated frame. The + symbol represents matrix multiplication and is used to define new frames in terms of an initial frame and a transform.

PAL:

```
%z        0  0  0  0  0  0;
%e        0  0 10  0  0  0;
%con     20 20 10  0  0  0;
%pal     10 10 10  0  0  0;
%app      0  0 +3  0  0  0;
%pos      1  0  0  0  0  0;
%row     -3  1  0  0  0  0;
tol   ::= e;
arm   ::= z+t6;
for i:= 1 to 3 do
   beg
      for  j:= 1 to 3 do
         beg
            mov c+a;
            mov c;
            gra;
            mov c+a;
            mov p+a;
            mov p;
            rel;
            mov p+a;
            p:=p+pos
         end
      p:=p+row
   end
```

**Figure 15.5** (Bonner) Palletizing with PAL.

```
MAPLE :

    /* MAPLE program to load pallets
       Declare data structures and initialize points */
          DCL  (conveyor,pallet,down)  FIXED POINT;
          DCL  (oneposition,tonextrow)  FIXED POINT;
          DCL  (ready,moving,i,j)  INTEGER;
          DCL  (palletflag,conveyorflag)  INTEGER;
          DCL  (done)  BOOLEAN;
          conveyor = PT(20,20,15)  IN LABXYZ;
          pallet = PT(10,10,15)  IN LABXYZ;
          down = PT(0,0,-3)  in LABXYZ;
          oneposition = pt(1,0,0)  IN LABXYZ;
          tonextrow = PT(-3,0,1)  IN LABXYZ;
          ready = 1;
          palletflag = 2;
          moving = 1;
          conveyorflag = 3;
          REFERENCE = HANDXYZ;
          done  := FALSE;
    /* Loop through pallet positions */
          WHILE NOT done DO;
          BEGIN
    /*    Make sure pallet is in place */
          WHILE STATUS (palletflag)<>ready DO; END;
          i  :=1;
          WHILE i<=3 DO;
            BEGIN
    /*        Make sure conveyor is no longer moving */
              WHILE STATUS (conveyorflag)=moving DO; END;
              j  :=1;
              WHILE j<=3 DO;
                BEGIN
    /*            Pick up block and place in pallet */
                  MOVE TO conveyor;
                  MOVE BY down;
                  UNLESS HIT DO; OPEN BY -.01; END
                  MOVE TO conveyor;
    /*            Ask user to restart conveyor */
                  OUTPUT TO TERMINAL 'Restart conveyor';
                  WHILE STATUS (conveyor)><moving DO; END;
                  MOVE TO pallet;
                  MOVE BY down;
                  OPEN BY 10;
                  MOVE TO pallet;
    /*            Move to new pallet position */
                  j  := j+1;
                  pallet  := pallet TRANSLATED oneposition
                END;
    /*        Move to new pallet row */
              i:=i+1;
              pallet  := pallet TRANSLATED tonextrow
            END;
    /*    Prompt user to replace pallet */
          OUTPUT TO TERMINAL 'Replace Pallet';
          WHILE STATUS(palletflag)=ready DO; END;
          END;
```

**Figure 15.6**   (Bonner) Palletizing with MAPLE.

## Example 15.4. Palletizing with MAPLE

The MAPLE programming language has user callable subroutines, and complex data and control structures. Because it has English-like command syntax, it is fairly easy to read even by someone who is unfamiliar with the language. The example MAPLE program shown in Figure 15.6 is very similar in structure to the PAL version, but is somewhat more readable.

## Example 15.5 Palletizing with AML/E

AML is the principal robot programming language delivered by IBM in support of its newest line of robots. AML is particularly attractive since it is being generated to run on the IBM line of personal computers. This example is in AML/E, a restricted subset of AML.

The example in Figure 15.7 makes extensive use of subroutines to implement the functions of move, pick, and place and uses the ITERATE function to conveniently sequence the robot through the pallet positions. The positions themselves are defined here as constants, P11-P33.

## Example 15.6. Palletizing with AL

The AL language allows callable procedures with parameter passing and the definition of complex data structures. All variables must be declared and can be assigned self-descriptive names. All decision points are described in terms of frames. In Figure 15.8, the NILROT vector is used to indicate that there are no desired changes in rotation. The AL program proceeds in the same manner as the one written in

```
CONV:   NEW PT( - - - );
P11:    NEW PT( - - - );
P12:    NEW PT( - - - );
P21:    NEW PT( - - - );
P22:    NEW PT( - - - );
P23:    NEW PT( - - - );
P31:    NEW PT( - - - );
P32:    NEW PT( - - - );
P33:    NEW PT( - - - );


PICK:   SUBR;
        DOWN;DELAY(1.5);GRASP;DELAY(1.5);UP;
END;

PLACE:  SUBR;
        DOWN;DELAY(1.5);GRASP;DELAY(1.5);UP;

END;

MOVE:SUBR(GET,PUT);
        PMOVE(GET);PICK;
        PMOVE(PUT);PLACE;

END;

ITERATE('MOVE',CONV,<P11,P12,P13,P21,P22,P23,P31,P32,P33>;

END;
```

**Figure 15.7**  Palletizing with AML/E.

```
            PROGRAM loadpallets;
            BEGIN
   AL :     ( Declare Variables )
              BOOLEAN done;
              EVENT palletthere, palletgone, nextpallet;
              EVENT conveyorstopped, conveyorgo;
              SCALAR i,j;
              VECTOR con,pal,mid,pos,row;
              FRAME conveyor,pallet,midpoint;
              TRANS newposition,newrow;
            (  Initialize variables  )
              done <-- FALSE;
              con <-- VECTOR(20,20,10) ;
              conveyor <--FRAME(NILROT,con) ;
              mid <-- VECTOR(15,15,15) ;
              midpoint <-- FRAME(NILROT,mid) ;
              pal <-- VECTOR(10,10,10);
              pallet <-o- FRAME(NILROT,pal) ;
              pos <-- VECTOR(1,0,0) ;
              newposition <--FRAME(NILROT,pos) ;
              row <-- VECTOR(-3,1,0) ;
              newrow <--FRAME(NILROT,row) ;
            ( Loop through pallets )
              WHILE NOT done DO
                BEGIN
                ( wait until empty pallet is positioned )
                  WAIT palletthere
                (  loop through pallet positions )
                  FOR i<--1 STEP 1 UNTIL 3 DO
                    BEGIN
                      FOR j<--1 STEP 1 UNTIL 3 DO
                        BEGIN
                ( wait for block to reach pick up point )
                          WAIT conveyorstopped;
                ( place block in pallet )
                          place(conveyor,pallet,midpoint);
                ( set new pallet position )
                          pallet <-- pallet+newposition
                        END;
                ( set new pallet row )
                      pallet <-- pallet+newrow
                    END;
                ( signal to replace pallet and wait until removed )
                      SIGNAL nextpallet+newrow
                      WAIT palletgone
                    END;
              END;
              ( Procedure to pick up block and place into pallet )
            PROCEDURE place(conveyor,pallet,midpoint);
              BEGIN
            ( Initialize deproach vectors )
                VECTOR down,up
                down <-- VECTOR(-3,0,0);
                Up <-- VECTOR(3,0,0);
            ( Pick up block and restart conveyor )
                MOVE bhand TO conveyor WITH APPROACH=down;
                CENTER bhand;
                MOVE bhand TO midpoint WITH DEPART=up;
                SIGNAL conveyorgo;
            ( Place block in pallet)
                MOVE bhand TO pallet WITH APPROACH=down;
                OPEN bhand TO 10.0;
                MOVE bhand TO midpoint WITH DEPART=up;
              END;
```

Figure 15.8    (Bonner) Palletizing with AL.

MAPLE. Although AL does not resemble English quite as closely as MAPLE, it has excellent facilities for interaction with external devices.

## 15.5 TASK-ORIENTED LEVEL

The task-oriented level of robot programming is a yet unachieved concept of a language that conceals low-level aids like sensors and coordinate transformations from the user. AUTOPASS (Lieberman and Wesley, 1977) is a language proposed by IBM to meet these criteria. It is designed to resemble instructions that might be given to a human assembly worker.

AUTOPASS uses high-level commands such as PLACE object1 ON object2. Execution of this command involves finding and identifying object1 and object2, determining a pickup point and approach vector for object1, moving to pick up object1, deciding where on object2 to place object1, placing object1 on object2, and remembering the new relationship between them. AUTOPASS requires a sophisticated world modeling system to keep track of objects. Ideally, this would include both visual location and identification of parts and use of tactile sensors for help in exact positioning.

AUTOPASS commands are divided into four types: state change statements (such as PLACE), tool statements (such as OPERATE), fastener statements (such as RIVET), and miscellaneous statements (such as VERIFY). The high level of the statements, however, can lead to ambiguities between the user's intended actions and the robot's interpretations of those statements.

To resolve these ambiguities, the proposed AUTOPASS system is designed so that program debugging will proceed interactively with the user. Commands are interpreted into lower-level code, and the user must verify the validity of the computer's interpretation. The user can alter any segment he or she wishes, and the compiler can question the user about any ambiguities in the AUTOPASS code.

AUTOPASS remains unimplemented and still does not solve the problems of collision avoidance and emergency decision making (Lozano-Pérez, 1980).

### Example 15.7. Palletizing with AUTOPASS

The AUTOPASS language depends heavily on a dynamic world model of the robot environment. The routine in Figure 15.9 is obviously very English-like and easy to understand, but it assumes that a great deal of information will be provided by the world modeling system. The Boolean functions emptypalletthere and holeinpallet-free are questions to the vision system asking whether the condition is true. The world model must be able to keep track of where the blocks are on the conveyor

```
AUTOPASS :

        WHILE emptypalletthere DO
        BEGIN
            WHILE holeinpalletfree DO
            BEGIN
                INSERT block IN holeinpallet;
            END;
            replacepallet
        END;
```

**Figure 15.9**   (Bonner) Palletizing with AUTOPASS.

(there is no need for a predesignated pickup point). It must also be able to locate and identify a holeinpallet, change the state after the hole has been filled, and decide when the pallet is full or empty. In addition it must be able to decide on approach vectors and grasping configurations.

## 15.6 CONCLUSION

In this chapter, we have attempted to categorize robot programming languages by both structure and capabilities. There exists already quite a variety in all categories, and the field is rapidly changing as new languages are developed and existing languages are supplemented with new features. In the future, we can anticipate that the level which we designate as *structured programming* will come to include controllers for most computer-controlled robots.

The development of such languages is a challenging problem in itself, since it entails real-time programming and concurrent processing, and requires a sophisticated understanding of the mathematical constructs developed in earlier chapters of this book.

## 15.7 REFERENCES

BAUMANN, E. Model-based vision and the MCL language. IEEE SMC Conference, Atlanta, October, 1981.

BONNER, S. A. Comparative study of robot programming languages. MS thesis. Rensselaer Polytechnic Institute, Troy, N.Y., 1983.

DARRINGER, J., and BLASGEN, M. "MAPLE: A high-level language for research in mechanical assembly." *IBM Research Report*, RC 5606. Yorktown Heights, N.Y. 1975.

General Electric. *Automation systems A12 assembly robot operator's manual.* Bridgeport, Conn., February 1982.

GROSSMAN, D. Programming of a computer-controlled industrial manipulator by guiding through the motions. *IBM Research Report*, RC 6393. Yorktown Heights, N.Y., 1977.

LIEBERMAN, L., and WESLEY, M. AUTOPASS: An automatic programming system for computer controlled mechanical assembly. *IBM Journal of Research and Development*, July 1977.

LOZANO-PÉREZ, T. "Automatic planning of manipulator transfer movements (M.I.T. AI Lab). *Memo 606*. Cambridge, Mass. December 1980.

MC DONNELL DOUGLAS. MCL language definition. November 1980.

SAHID, M. Current status of the AL manipulator programming system. *Proceedings of the 10th International Symposium on Industrial Robots*, Milan, 1980.

—, and GOLDMAN, R. The AL user's manual, STAN-CS-79-718. Stanford University, Palo Alto, Calif., 1979.

TAKASE, K., PAUL, R., and BERG, E. A structural approach to robot programming and teaching. *IEEE Transactions on Systems, Man, and Cybernetics*, SMC-11, no. 4, pp. 274–280, April 1981.

Unimation, Inc. User's guide to VAL. Danbury, Ct., 1979.

# INDEX